Every Other Thursday

MASS AWIS 10th Anniversary

September 2014

Every Other Thursday

STORIES AND
STRATEGIES FROM
SUCCESSFUL WOMEN
SCIENTISTS

Ellen Daniell

YALE UNIVERSITY PRESS   NEW HAVEN AND LONDON

Designed by Rebecca Gibb. Set in Janson type by Integrated Publishing Solutions. Printed in the United States of America.

The Library of Congress has cataloged the hardcover edition as follows:
Daniell, Ellen, 1947–
Every other Thursday: Stories and strategies from successful women scientists / Ellen Daniell.
p.   cm.
Includes bibliographical references.
ISBN 13: 978-0-300-11323-5 (alk. paper)
ISBN 10: 0-300-11323-4 (alk. paper)
1. Women in science. 2. Women scientists. 3. Science—Vocational guidance.
I. Title.
Q130.D36 2006
305.43'5—dc22   2005018885

ISBN 978-0-300-51084-3 (pbk. : alk. paper)

A catalogue record for this book is available from the British Library.

The paper in this book meets the guidelines for permanence and durability of the Committee on Production Guidelines for Book Longevity of the Council on Library Resources.

10 9 8 7 6 5 4 3 2 1

To Group — Christine, Beth, Helen, Mimi, Suzanne, Judith, and Carol — the heart, soul, and cast of characters of this book.

And to my mother, Winifred Marvin Daniell, whose love and support in all my endeavors have been the greatest constant in my life.

Contents

PREFACE xi

INTRODUCTION: WORKING TOWARD DIVERSITY xv

A NOTE ON CONFIDENTIALITY AND TERMINOLOGY xxv

Part 1 Why Group?

1 Every Other Thursday: A Meeting of Group 3

2 Evolution: How Each of Us Came to Group 17

3 Facing Disaster: Ellen's Story 35

Contents

Part 2 Group Work

4 Accepting . . . Liking . . . Celebrating: Appreciating
Ourselves and Being Entitled to Success 71

5 A Serious Mind and a Light Heart: Respecting
Instinct and Personal Goals 82

6 Off Balance and Out of Control: Managing Time
and Establishing Equilibrium 91

7 Flying Furniture: Choice and Change 101

8 Best Friends, Harshest Critics: Working
with Other Women 108

9 Life Is a Limited Resource: Taking Care of Ourselves 118

10 Permission to Feel: Being Professional Does
Not Mean Turning Off All Emotion 127

11 Boss, Mother, Friend, Role Model: Working
with Students and Employees 138

12 Putting It Out There: Writing and Giving Talks 149

13 Nobody Taught Us This in School: Institutional
Politics and Strategy 161

Contents

14 Anticipating Changes: Growing Older with Grace 177

15 Going Home: Interactions with Spouse,
Partner, Mother, and Children 184

Part 3 A Group of One's Own

16 Pigs, Contracts, and Strokes: Group
Process and History 201

17 Maintenance and Repair: Working to
Keep Group Working 227

Part 4 Epilogue

18 Another Change of Direction: Letting Go
and Moving On 239

Biographies of Group Members 253

NOTES 259

FURTHER READING 263

ACKNOWLEDGMENTS 266

Preface

- *If I'm not busy every second of every day, it seems that I'm not working hard enough.*
- *Maybe having a fulfilling personal life is incompatible with a successful career.*
- *I feel like I'm an emotional cafeteria, responding to what others want.*
- *I feel responsible for everything but have no power to change anything.*

These perceptions come from members of a problem-solving group that aims to empower professionals by providing practical and emotional support. Having been a member of this group for more than twenty-five years, I have written this book to encour-

age others to try a group approach in countering stress and isolation in professional life. Discussions with people in diverse careers
have convinced me that many of the fears, concerns, and strategies of our group are widely applicable; I have set out to share
them in these pages. In addition to examples and illustrations, I
have included practical guidelines.

The objective of Group, as we call it, is cooperation in the
competitive world. Group members seek both practical solutions for specific problems (such as dealing with a difficult boss
or employee) and broader perspective on our lives. Group helps
counter the all-too-common experience of professional life as a
combat zone in which nobody seems to be on your side. (The
phrase "swimming with sharks" is often used to describe life in
the business, legal, and academic worlds.) Anyone who feels isolated in a professional or competitive setting or who wants honest feedback can benefit from a group, a safe testing ground where
everyone is on your side. A critical message of this book is that intimacy and reliance on others for encouragement and advice is a
source of empowerment, not a sign of weakness.

Every Other Thursday is based on the professional experiences
of seven women. Although most of them work in an academic
environment and began their careers twenty-five to thirty years
ago, their messages remain relevant. The struggle to achieve
gender and racial equality, for example, is an ongoing concern in
academic and business settings. I do not attempt an in-depth ex-

ploration of this issue, but the Introduction summarizes extensive data that demonstrate its immediacy.

This book is not only for women. Men who have read it have urged me to promote its message to males as well. The point is that men *or* women who find themselves part of a minority in any professional setting share many of the psychological and emotional challenges described here. Group problem solving is valuable in facing a variety of challenges—from embarking on a new career to dealing with conflict to preparing for retirement.

The process I describe is not about getting advice on how to behave, manage one's time, or be a good boss; rather, it is about support, discussion, and practical application of common experience to individual problems. The group approach derives its strength from numbers, because of both the diversity of feedback and the experience of sharing others' problems.

This is neither a memoir nor a history of our group. The story of my own life as a scientist, professor, businessperson, and writer exemplifies how Group works: its members helped me navigate a major career setback and then identify and pursue my professional dreams. The lives of other members are seen through their work in Group meetings.

Our years of discussion in Group have revealed a number of themes, which include recognizing personal achievements and taking credit for them; finding balance in managing busy lives;

and making choices with a belief in our right to make them. One chapter addresses the rewards and complexities of relationships among women in the workplace. Others deal with specific professional challenges such as writing, public speaking, and institutional politics. Issues of particular concern to women (but relevant to men as well) include the need to acknowledge and honor their feelings and the struggle to incorporate our families and careers in one life. Part 3 details the rules and procedures of our group, explains why we found them important, and provides guidelines for readers interested in starting a group. The Further Reading section lists works on groups and group therapy for those who wish to pursue the topic.

In this book I encourage the reader to look to others for support in professional and personal endeavors. I have witnessed women successfully help one another work through dilemmas and celebrate accomplishments through the dynamics of group interaction. This book is about the power of *not* trying to go it alone.

Introduction: Working Toward Diversity

Every Other Thursday focuses on women scientists who created an association to help one another through the complexities and stresses of their competitive careers. In the 1970s, when the group started, choosing such a career frequently meant being the first or only woman in a department or research organization. Today, as the numbers of women in science and engineering have increased, such extreme isolation and obvious groundbreaking is less likely. Are the struggles of these women and the lessons they learned therefore only of historical interest today? The answer is an unequivocal no, for several reasons. First, analysis of the numbers of women advancing to top levels in their careers shows that a gender differential persists. In-depth studies on career progression and the environment women experience in the workplace demonstrate that very real barriers remain. Furthermore, it is clear that members of racial and ethnic minorities share many of the effects of isolation that women have experienced and with which they still struggle.

Despite progress toward equity, important disparities still exist. Although this volume attempts neither to be an authoritative source of data nor to determine causes of discrimination and underrepresentation of women in science, key studies provide a current context for the message of this book.

A primary source of the power of the group process is that groups bring together people with shared concerns who might otherwise be isolated from one another. Data on the representation of women illustrate this point. (The numbers for other minorities are still more striking.) In 1999, women earned 36 percent of doctoral degrees granted in science and engineering (up from 8 percent in 1973). Yet only 26 percent of employed doctoral scientists and engineers were women. Of those employed in four-year colleges and universities, 29 percent of the doctoral scientists and engineers were women. Within academia, the percentages decrease when tenured or full professors are considered and when the most prestigious research institutions are examined. A publication entitled "A National Analysis of Diversity in Science and Engineering Faculties at Research Universities" looks at the representation of women in fifty top research universities.[1] The authors discovered that 7.6 percent of full professors in chemistry and 14.8 percent in biological sciences were women. The numbers were even lower for mathematics, physics, and all branches of engineering. In sociology and psychology, the fields having the highest percentage of women faculty, there was still a striking contrast between assistant professors (52 and 45 percent, respectively) and full professors (14 percent in both fields). Abigail Stewart, project director at the Institute for Research on Women and Gender at the University of Michigan, has pointed out that "many smart motivated women have cited isolation and marginalization as reasons for moving out of science and engineering at major research institutions." To extend the point to fields outside academia, it is interesting to note that as of 2001 only four of the CEOs of Fortune 500 companies were women.[2]

That women have lagged behind men in their representation in the science and engineering workforce is of much concern to organizations that seek to foster scientific achievement. In researching the issue, I discovered that the situation is worse than I had imagined.

The administration and faculty at Massachusetts Institute of Technology (MIT) performed the first and most famous in-depth study on the status of women faculty within a particular institution. A group of senior women on the faculty had gathered preliminary evidence that they had less laboratory space, less access to research funding, and lower salaries than their male counterparts. In addition, they were infrequently represented on committees that made decisions about hiring and research funding. MIT's administration responded by researching the charges, finding that they were accurate, and taking steps to correct the inequities. The abstract to their report is an excellent description of the issues that still confront women scientists and analysis of why they went unrecognized by administration as well as by the women themselves.[3]

In 1995 the Dean of Science established a Committee to analyze the status of women faculty in the six departments in the School of Science. The Committee submitted a report of its findings in August 1996 and amended reports in 1997 and 1998. The Committee discovered that junior women faculty feel well supported within their departments and most do not believe that gender bias will impact their careers. Junior women faculty believe, however, that family-work conflicts may impact their careers differently from those of their male colleagues. In contrast to junior women, many tenured women faculty feel marginalized and excluded from a significant role in their departments. Marginalization increases as women progress through their careers at MIT. Examination of data revealed that marginalization was often accompanied by differences in salary, space, awards, resources, and response to outside

offers between men and women faculty with women receiving less despite professional accomplishments equal to
those of their male colleagues. An important finding was
that this pattern repeats itself in successive generations of
women faculty. The Committee found that, as of 1994, the
percent of women faculty in the School of Science (8%)
had not changed significantly for at least 10 and probably
20 years. The Committee made recommendations for improving the status of senior women faculty, addressing the
family-work conflict for junior women faculty, and increasing the number of women faculty. The Dean of Science took immediate actions to effect change and these
have already resulted in highly significant progress including an increase in the number of women faculty. This col-
laboration of faculty and administration could serve as a
model for increasing the participation of women, and also
of under-represented minorities, on the faculty of other
Schools at MIT. This is an important initiative since, even
with continued effort of this magnitude, the inclusion of
substantial numbers of women on the Science and Engineering faculties of MIT will probably not occur during
the professional lives of our current undergraduate students. The inclusion of significant numbers of minority
faculty will lag for even longer because of the additional
problem of a shortage of minority students in the pipeline.

A study of the persistent gender gap in the science and engineering workforce by the National Research Council (NRC) concluded that "while women are now visible in fields in which they were virtually absent 25 years ago . . . they continue to cast a very small shadow." Finding that "at any age women lag behind their male colleagues in career advancement," the report discusses conditions that slow or interrupt careers, childbearing and rearing being the most obvious but not the only factors. An analysis of salaries shows that a gap also persists there, even when the data are broken down to compare only "groups that are as comparable as possible." For example, there is still an 11 percent gap in salary between men and women who are full professors. Because of conflicts with family responsibilities, women may choose lower-paying jobs that have greater nonsalary rewards such as time flexibility.[4]

Discussing the effects on career outcomes of being in a minority, the authors of the NRC study (*Scarcity to Visibility*) comment, "A given field . . . may need at least a minimum number of women before these women attain a critical mass whereby they are no longer viewed as an oddity. Having a critical mass can minimize socialization difficulties otherwise encountered in a male-dominated environment." Many studies noting the attrition of women from science at every stage of education and promotion suggest the lack of role models as one cause. (In the case of ethnic minorities the disparity is even greater, so that a Black or Hispanic individual, male or female, is very likely to go through training without a single role model of his or her ethnicity.) Dr. Debra Rolison of the Naval Research Laboratory postulated that "a very plausible case can be made that academic departments are an unhealthy—even hostile—environment for women." Professor Janet Osteryoung, director of the Division of Chemistry at the National Science Foundation, described the reaction of young women to what they see: "Women who are eligible for faculty positions have earned a Ph.D. in a chemistry department. They have absorbed the tone of that environment . . . and have decided they don't want any more of it."[5]

On January 14, 2005, Lawrence Summers, president of Harvard University, spoke at a conference on women and minorities in science organized by the National Bureau of Economic Research (NBER). He cited studies showing that more boys than girls score at the high and low ends of standardized math and science tests and suggested that this could explain the larger number of men at the top of the professional ladder in these fields. He specifically stated his belief that this difference in "availability of aptitude at the high end" had a greater impact than social factors and discrimination on the underrepresentation of women.[6]

Among the flaws in Summers' analysis were the lack of evidence that genetic factors are responsible for the difference in test scores and the substantial evidence that test-score differences do *not* correlate well with success in the professions. Participants in the conference noted that they had been listening throughout the day to complex and careful studies demonstrating the role of discrimination and socialization. In this context, Summers' comments seemed careless and gratuitous in light of the chilling effect they would doubtless have on efforts to combat discrimination, coming from someone in his prominent position. Dr. Yu Xie, a sociologist from the University of Michigan (present at the conference), whose work Dr. Summers had cited, said that his data had been misinterpreted. Xie stated that the inference that the underrepresentation of women in the top ranks of science and math could be due to differences in ability was an illogical leap. Research shows that even women at the highest levels of mathematical achievement are less likely than their male peers to pursue math and science careers. Xie and others also noted that over the past few decades there has been a steady increase in women's participation in the sciences and engineering. Because the genetic pool has not changed in that time, factors like improved educational conditions for women and efforts to reduce discrimination are much likelier explanations.[7]

Summers introduced his remarks by saying that he had made "an effort to think in a very serious way about" the lack of women in the higher echelons of math and science. Citing areas that might be studied in the same way, he observed that "white men are very substantially underrepresented in the National Basketball Association; and that Jews are very substantially underrepresented in farming and in agriculture." In a letter of apology written a month later, he said, "My January remarks substantially understated the impact of socialization and discrimination, including implicit attitudes—patterns of thought to which all of us are unconsciously subject. The issue of gender difference is far more complex than comes through in my comments, and my remarks about variability went beyond what the research has established."[8]

It was perhaps most unsettling that those who protested Summers' comments were accused of trying to stifle scientific debate and of demanding politically correct speech that precluded suggestions of genetic differences. An organizer of the conference suggested that the critics were activists whose sensibilities might be at odds with intellectual debate. In this way the important discussion of prejudice leading to scientifically invalid claims was partly submerged in an uproar about free speech that missed the point at hand.[9]

The extensive discussion following Summers' remarks did appear to have a positive result—the mobilization of scientists and sociologists to set the record straight and to communicate concerns about the suggestion that underrepresentation of certain groups had inevitable natural causes. This discussion will benefit not only women but also other groups who do not find science and engineering fields hospitable. A letter to *Science* magazine signed by more than eighty scientists concludes: "We must continue to address the multitude of small and subtle ways in which people of all kinds are discouraged from pursuing interest in scientific and technical fields. Society benefits most when we take full advantage of the scientific and technical talent among us. It is time to create a broader awareness of those proven and effective means, including institutional policies and practices, that enable women and other underrepresented groups to step beyond the historical barriers in science and engineering."[10]

In this context, the formation of problem-solving groups could contribute significantly to overcoming barriers to participation in any profession. The stories of the women in *Every Other Thursday* and the strategies they developed to achieve success and balance attest to the effectiveness of the group process.

A Note on Confidentiality and Terminology

The issue of confidentiality in writing this book was critical. Many writers who draw on personal experience are troubled by the possibility of compromising the privacy of their friends. In writing about Group, where everything discussed in our meetings is confidential by formal agreement, the need to respect confidences goes beyond the rules of common courtesy and the dictates of friendship. All the members of Group have supported the idea and the substance of this book, and each woman has had the opportunity to review and approve my discussions of her work. I sometimes use an example without specifying who is working, usually to protect a student, colleague, or family member—and only if I feel the story is effective without attribution. The chapters on my own life are a refuge from the burden of confidentiality, the writing I could do with only my own privacy to consider.

Certain frequently used terms may be unfamiliar to readers. For convenience, I provide brief definitions here. (They are explained at length in Part 3, Chapter 16.)

Pig: a negative self-perception, an external judgment that one assumes and uses to defeat oneself. In our Group work, we sometimes identify pigs by naming them.

Contract: a concise formulation of objectives, either immediate or long-range, to solve a problem or reach a goal. A contract should be "do-able," recognizing that large problems are made up of lots of small ones.

Stroke: a positive statement about someone else. At the end of each meeting, time is dedicated to giving (and accepting) strokes.

Part 1 Why Group?

1 *Every Other Thursday*

A MEETING OF GROUP

It's a Thursday evening in November and my turn to host Group! I leave work in time to stop at a deli on the way home. I wish I had time to cook, the way Helen does, but she's retired and I'm not . . . yet . . . so the deli it is. I lay a fire; that at least is a homey touch. I try to slow down a little, to think about what I want to talk about this evening, wishing I had reserved a few minutes to write some thoughts in my notebook. I have resolved to do advance preparation for my Group work, but I don't stick to my resolve as often as I'd like.

In addition to the pleasure of entertaining these women, I relish having Group at my house because of the luxury of not having to drive home afterward. The meetings often run until midnight, and I always feel a conflict between the dread of being exhausted the next day and my enjoyment of the discussion and laughter.

I put water to boil and get out tea and coffee. Do I need regular coffee for anyone? No, even the last holdouts have turned to decaffeinated beverages in the evening. The doorbell rings at 7:30, and Suzanne is on the doorstep, tall, red-haired, and elegant. "Hi. I allowed extra time in case traffic was bad, but it wasn't!" These moments before the meeting begins are a time to catch up on news with the early arrivers, before the structured work begins. I pull her into the kitchen while I continue to organize. "How are you? How is the family?" As usual, Suzanne's face lights up as she talks about her husband and children. "Maria is applying to colleges in the East, and I'm already thinking about how much I'm going to miss her." Maria is the youngest among the children of Group members, the last Group kid at home. The news continues. "Kit's fine. He just sent a wonderful e-mail from Mongolia, but I worry about him anyway. Arthur and Nancy are having a rough time looking for two academic jobs in the same place."

Christine arrives next. I hadn't been sure she was coming. She's missed several recent meetings owing to a complicated travel schedule, so I'm especially delighted to see her. I also crave reassurance that she is well, because it has been only a few months since she completed chemotherapy for breast cancer. I give her an enthusiastic hug. The doorbell rings again; Judith, Helen, and Mimi have carpooled from Berkeley.

Mimi's waist-length hair is damp; she has squeezed in a run with a colleague after work. "I'm impressed," I say; "you're doing the important things to take care of your body." "How was your trip?" asks Helen, always the best at keeping track of what we're all doing. "The time with my mother was great," I answer, adding a few less enthusiastic words about the business part of the trip. I'm trying to talk to everyone at once, pouting with Judith that we haven't managed to schedule a walk together in a month of trying. "We'll do it after Thanksgiving," she promises.

Only Carol is missing, but she had called earlier to say she might be late. I'm facilitator as well as host, so it's my responsibility to get the meeting started at eight. I start pouring coffee as a signal to begin. Everyone chooses a seat, paying attention to their back needs. Helen sits on the floor. Judith chooses the sofa. Christine joins her, then thinks better of it and takes the antique wooden rocker. I bring straight-backed chairs from the dining room for Suzanne and me. People get out their notebooks. There is preliminary chatter: "Do you have some paper? I forgot my notebook." "I need a pen or pencil." "Who wants tea, who wants coffee?" "It *is* decaf isn't it?"

"Let's get started." I take my notebook out and call us to order. "Does anyone have any feelings they'd like to share?" Suzanne responds, "I'm exhausted and glad I made it." Mimi confesses, "I want to sleep." I offer, "Shall we go straight to strokes and food?" Then I say seriously, "I'm so happy that everyone is here."

"Who would like to work?" I ask. Each person in the circle states how much time she would like. I make notations: Mimi, fifteen minutes; Christine, ten minutes; etc. Tonight everyone asks for ten or fifteen minutes, suggesting that we all have issues to work on, but perhaps not terribly heavy ones. Probably everyone is recognizing the time constraints suggested by having a full Group. When fewer members attend, we miss the wisdom of those who are absent, but we can be more relaxed about time without fear of running too late. Tonight, as facilitator, I'll have to pay particular attention to letting each person know when her requested time is up.

Carol arrives at 8:05. "I'm sorry. I had a student committee that didn't end until seven." She sinks into the remaining spot on the sofa with a sigh. "We've just finished saying how much time we want. Do you want to work?" She shrugs, "Oh, I don't know. I guess ten minutes." By the time Carol joined Group, the rest of us had been together over ten years and had gotten a little lax about keeping to time, so she never has developed the habit of trying to estimate her time. She is also the only one who doesn't keep a notebook to record thoughts and contracts.

I look around the room and get a rush of appreciation. I feel better about myself in the presence of these women, and I expect that in the course of the evening I will feel more in charge of my life and will gain clarity about the issues I plan to raise. We have been meeting for twenty years, yet every meeting is an adventure. I'll learn something new about life, or myself, or a facet of someone's character, or all of the above.

Next I ask, "Who wants to work first?" I try to catch someone's eye and chuckle as people suddenly begin to study their feet or gaze into the fire to avoid starting. But tonight Suzanne says, "I can go first." Everyone sits up a little and looks at her. "I'm having the winter blues. I am worried about all things. I'm actually very happy in the lab, and good things are happening, but the soft money problem is really bothering me. I feel like I'm up for tenure every year forever." (As an investigator in a "soft money" institution, Suzanne has to write grant requests to cover her own salary, in contrast to university faculty, who have nine months of their salary paid by the institution.)

With this background about her general state of mind, Suzanne moves on to a specific issue. A message from her former postdoctoral adviser telling her that she is "one of the best of her generation of scientists" in their field has produced anger instead of the delight that seems to be the expected response to such a compliment. "Why did he *never* give me encouragement when I was young, struggling, and in much greater need of it?" We all understand the dilemma, because Suzanne has worked on her complicated professional relationship with this man before.

Someone validates her response, saying it's appropriate to be angry with a mentor who withholds deserved praise. "Now that he has come through, you realize how much he has been withholding." Several people suggest ways Suzanne might respond; she decides she'd like to tell him, "It would have helped me enormously if you had said that earlier." Glancing at my watch, I realize I've been so engrossed in the discussion that we've gone over the fifteen minutes Suzanne requested. "Your time is up; would you like more time?" Suzanne thinks and shakes her head, "No, this has been helpful, and I'm done." Instead of moving on, I say one more thing. "While you're dealing with these reactions, don't forget to take some pride and pleasure in the fact that he feels that way." She agrees, but without enthusiasm. She may appreciate his compliment in the future, when she has dealt with her resentment, but not yet. Judith and Helen have another item, in response to Suzanne's blues, which she's suffered from in other years. They have read of indoor lamps that are supposed to combat depression arising from insufficient daylight. She agrees to look into the product.

"I'll go next, before I get too tired," Mimi says. "I have one more lecture to go in my big course. After that I can work on my mini-contract from last time to get down to choosing anti-stress strategies." In the previous meeting she had told us about attending a workshop on dealing with stress, and someone had suggested the affirmation "Life's events happen. How we feel about it is up to us." Mimi goes on: "I feel incompetent because I'm so unable to cope, unable to keep up with all the demands that are made on me day after day. I'm focusing only on the people and things that are mad at me, so I'm surprised when someone isn't. Yesterday someone did something nice to help me out and I burst into tears." She ends her work by reaffirming the contract about strategies to reduce stress, and adds, referring to her upcoming sabbatical, "I'm hanging on until we go away in January."

Judith, connecting with Mimi's work on stress and overload, asks to go next. "I have been saying no to invitations and re- quests, creating space in my life. Now I'm compulsively filling up the space, dotting *i*'s and crossing *t*'s. I'm falling back on old habits. My need to be somebody is tied up in work, and work feels empty." She sees younger scientists going through the anx- iety and pressure to produce that she once experienced, and she feels exhausted, somehow still caught up in those pressures, even though she thought she had gone beyond them. Moving on to other difficult work, she says, "I often feel like my feelings are frozen. I have so many ways of avoiding getting through to my feelings." She makes a list: "raging, going numb, working in- sanely hard." Then she exclaims, "I'm stuck. Stuck in Group, stuck in life, stuck, stuck, stuck!" There's a pause. No one is sure how to help. Mimi breaks the ice with, "I'm depressed, Suzanne's frantic, and Judith's STUCK." The laughter feels good, and Judith gets unstuck enough to proceed. "I'm frantic about filling the gaps I've created." For the short term she proposes, "I will do nothing for an hour a day and watch what comes up." She articulates a more long-term contract, too: "I will honor my battle." Helen adds a suggestion: "Learn to love your gaps."

I look at Carol, Christine, and Helen to see who's interested in going next, and Christine speaks up. "I'm in meltdown. I've been trying to work on my organizational skills and my therapist said, 'You don't need any more organizational skills, you need time to organize.' I have plenty of help [in the lab, in her home]. I need to take time to tell them what to do." I think what a wonderful raconteur Christine is, as she describes a scene from her life as a metaphor. "John and I figured out a schedule for our lives over the next six months. We figured out how to do it on the computer, we finally got it done, and the computer crashed and erased it all. I feel like I've surrendered to a higher power." Getting down to specifics, she makes a contract to make time for necessary discussions with staff and students. Looking at her watch before I check mine, she says she's finished and adds with dramatic portent, "But this work will continue."

Carol says, "I want to talk about some emotional issues, and I'm amused that I'm *not* talking about all the lab problems. I meant to, but I figured out the most critical answers as I was planning how to tell you about them." She talks a little about family issues. Her mother, Mollie, aged eighty-five, moved in with her in the early summer, and they are getting used to each other. "Mom has to have things a certain way, and she has to or-ganize everyone to help it be that way. I, in contrast, will never organize other people." Carol, reflecting on how Mollie's style differed from that of Carol's husband, muses, "He also wanted things just so, but he would do them himself, not in a compulsive way, and it became fun to do them together." Carol doesn't really want feedback on this. She's using Group to voice explanations and emotions she's working out for herself.

Carol turns to the specter of the upcoming lectures she will have to give as part of "BioReg" (Biological Regulatory Mecha-nisms), a legendary course at the University of California at San Francisco taught by a team of high-powered faculty. "I still feel like the new kid on the block. This course demands perfection that I can't deliver." Christine, for whom poor teaching evalua-tions in that course were a major source of angst and depression in the past, is approaching the course in a different way. "I'm using it to think about science broadly and *my* science in particular. It's a totally different attitude, not terror." For Carol, it helps just to hear how hard it was for Christine and how lecturing, once diffi-cult for all of us, had become easier with time. Also, she tells us later, describing her fears helped her organize in her mind what she needs to do to prepare.

Helen reports that she has new hearing aids. "They are won-derful, and Health Net Senior Services paid for it all. I'm already hearing the difference." We all feel relieved, knowing that the old devices had been uncomfortable and that Helen had feared the higher cost of new technology. She has been thinking about her fears. "I spend time being fearful, but also time being fearless. Sometimes I don't speak out because of being fearful, but at other times because it feels right to be in the background." Encouraged by Group to define the fears, she listed "fear of saying something inappropriate, fear of not having enough money (but I have also taken risks with regard to money), fear of not marrying again [she pauses to question whether that is really a fear], fear of Death." Then she listed the brave things she had done: "adopting children, leaving my husband, buying a house alone, quitting my job." She's been enjoying her first grandchild, and her face becomes even more animated as she tells us about Salina. "She's learned to talk now and loves to sing. Yesterday I listened to her belt out a song, and with a little difficulty figured out it was 'Can You Feel the Love Tonight?' It's from *Lion King*. She probably doesn't have any idea what the words mean, but she has the video and she can sing it! I laughed and laughed."

I go last, as the facilitator usually does, asking that someone else keep track of my time. I am planning to quit my job, which has created a problem. "Now that I know I'm going to leave Roche, I've become extremely impatient with its shortcomings." Letting loose the feelings I have been suppressing, I rant about "idiotic decisions and choices." I'm concerned that my irritation will show and prevent me from accomplishing what I want to be-

Group Is . . .

- Women who came together to discuss professional concerns and have become confidantes and friends, continuing to meet for more than twenty years.
- Commitment to cooperative action in a competitive world.
- A forum for professional problem solving.
- A sounding board, a reference point, a source of perspective and challenge to comfortably held views. "What would Group say?"
- A meeting every other week, a session to be scheduled, a calendar priority. "Are you going to Group Thursday?"
- A source of personal enrichment, acknowledgment and enhancement of personal power, an arena in which to recognize and renew our authentic selves.
- A celebration of life, letting it all in. "I can hardly wait to tell Group."
- Solace, a lifeline, a place where we can expect any fear or weakness to be met with compassion and where we are committed to compassion for others.
- A chance to help one another, to offer opinions and share experiences.
- Twenty-four years of history, influenced by former members who have contributed to its conception, organization, and evolution.
- Hard work . . . hearty laughter . . . welcome home.

fore I leave. Most of my work involves patient compromise with managers in other parts of the company and skillful negotiation with outsiders. I'm good at it because I usually stay calm and laugh a little. "I think I'm losing it." Everyone agrees that my response to the situation is natural. "You now have the luxury of being pissed. You can't let yourself be that pissed off at a place until you know you are not going to stay there." Someone adds, "You're evolving—or maybe revolving—out the revolving door." I go on to talk about other sources of preretirement anxiety, admitting that I feared not having external approval for my achievements. More immediately, I am worried about telling my terrific boss that I'm retiring. I'm not planning to do it for several months, but I'm already anxious. Group advises that I dedicate time to the details, considering the best and worst things that could happen. My contract is, "I will write the 'quitting' scene as a play and imagine it exactly as I would like it to go." This kind of role playing has been useful in the past.

Judith says, "Your time is up." I nod, return to my facilitator role, and ask, "Time for strokes and wine?" Everyone pitches in to slice bread, open wine, pour water. We are easy with one another in the kitchen, and this preparation is the transition into the less formal part of the evening. I have set the table with heirloom china that my mother has recently given me, reminding Group of its special significance. We sit down, talking about this and that, filling our plates and glasses. After ten minutes, Judith says, "I have a stroke for Suzanne." Everyone quiets down. Suzanne looks attentively at Judith, preparing to accept the stroke. "You confronted the complicated feelings that compliment aroused in you instead of feeling guilty that you weren't just proud and pleased." Suzanne says simply, "Thank you." This is the best way to respond to a stroke, although we may sometimes add a few words, as long as they are positive. Helen has a stroke for Mimi, "about recognizing and appreciating the helpfulness of others in the midst of your stress." Mimi looks surprised, and I wonder if she's thinking of protesting, but she follows good-stroke etiquette and does not demur. I give Christine a stroke "for the way you tell a story and make us laugh while getting *right* to the heart of things." Suzanne says, "I have a stroke for Carol, for the care and attention you are giving to establishing your life with your mother." Christine follows up on that: "I have a corollary stroke. That you are taking care of yourself *and* considering your own feelings as well as Mollie's." Carol beams and nods, absorbing the appreciation. Judith says, "A stroke for Helen for her feedback about learning to love my gaps," and then, with a mischievous look, "and a visual stroke for Ellen." I sit up straight and try not to preen. "For how you look tonight. Those earrings are gorgeous with your hair, and you look elegant and comfortable." I thank her and reflect on how good everyone looks to me. Lots of gray hair in varying styles, laugh lines intensifying by the year, and a sense of forthrightness. Stroke etiquette prohibits a response in the vein of, "Oh but not as beautiful as everyone else," and I focus silently on my sense of pleasure and comfort.

We slip back to general talk about our lives, families, and mutual friends, punctuated with an occasional remembered stroke. Mimi gives me one about a walk we took recently, "busting ass" on steep trails as I asked her questions about her research. She beat me to the punch with that stroke; it was a wonderful walk, and I respond that I was astonished and impressed how clearly she explained unfamiliar science to me. And so on. Judith says regretfully, "I've *really* got to go. Tomorrow morning is looking awfully close." "Ouch," says Christine. "Me too." Everyone checks her watch and sighs. There is great attention to hugs all around and a few promises to call with information or to set up a date for a lunch or a walk. Carol and Suzanne get into a conversation while we're clearing the table and stay a little longer than the others. I load the dishwasher, turn off the porch lights when the last car has pulled out, and sit for a moment with the last of my wine. I'm rejuvenated, full of new ideas, more confident, and weary.

2 *Evolution*

How did Group start and evolve to its present form? Each of the current members joined for different reasons and has benefited from the process in different ways. Invited several years ago to give a presentation at a national scientific meeting, we chose to appear as a panel, each woman telling her own story. I have transcribed these talks, which describe in very personal terms the terrors and obstacles that each person encountered in establishing a professional life and which demonstrate how useful Group has been in helping them overcome these challenges.[1]

Women in Cell Biology (WICB) and American Women in Science (AWIS) sponsored our presentation at the annual meeting of the American Society of Cell Biology in December 1994. We were introduced and spoke in the order in which we had joined Group. Christine Guthrie, a professor of biochemistry at University of California, San Francisco (UCSF), and a founding member, spoke first. I (at that time Director of Licensing at a biotechnology company) went next. Helen Wittmer had retired five years earlier from a position as administrative assistant to the chair in the Department of Molecular Biology at the University of California, Berkeley (UCB). Next in line was Suzanne McKee, associate director of Smith Kettlewell Eye Research Institute in San Francisco. Mimi Koehl was a professor in the Department of Integrative Biology at UCB. Judith Klinman was a professor of chemistry and also a member of the Molecular and Cell Biology department at UCB. Carol Gross, the newest member, was a professor of Tissue and Cell Biology and vice chair of the Micro-biology Department at UCSF.

Christine

Within three and a half years after my arrival at UCSF, a series of traumatic events, which included the death of a very treasured colleague, a pessimistic mid-career review, and others, precipi-tated a severe clinical depression. During and after my hospi-talization, I was amazed at the outpouring of response from coworkers, many validating their own painful insecurities in the academic environment. Convinced up until then that I was the only one who couldn't cope with the stress of competition and the absence of nurturing in the hierarchical bureaucracy, it was incredibly liberating to learn that so many other colleagues, even men at the very top of their professions, shared the fear that they, too, would any day be discovered for who they really were—inadequate poseurs. An outgrowth of these conversations was my introduction to a unique community in the Bay Area, a group of very smart and highly gifted therapists who promoted the notion that by pooling our shared experiences and collective wisdom we at the university could empower ourselves to succeed professionally within the existing academic structure while at the same time providing critical emotional support to one another. Thus was born Group, which in one form or another has been meeting twice a month for eighteen years.

Of course, there are many stories to tell—Group has been one of the longest sustained relationships in my life. But I prefer to highlight briefly three phases of work that I've done in Group, all of which have significantly changed my life. The first was during the early years, when I had to face the prospect of returning to my job after a six-month medical leave to confront an uphill struggle for tenure. Group helped me design successful scientific and political strategies, even constructing my tenure packet. Second, and surely as important, Group provided the crucial emotional support that convinced me I could make it and assured me that even if I didn't, there were meaningful alternatives. In 1980, three years after Group began, I received tenure.

The second is work of the middle years, when I struggled to

confront extremely negative teaching evaluations that deepened my already profound fears of performing in the classroom. Group labored long and mightily on this one, at one point resorting to a musical comedy routine for me to practice before facing an audience, the opening lyrics of which were "Howdy terror!" Probably few of us are ever completely satisfied with our teaching evaluations, much less our actual performance, but at least the fear has become manageable and allowed me to focus on the substance of the lectures.

The third is work of recent years, which involves making peace with personal sacrifices that I made in choosing to fulfill the consuming demands of a high-profile career. Once again, Group has risen to the challenge of this emotionally charged work, which has become very much a shared theme. Let me end by simply saying that it is perhaps the perfect irony that Group has become the family that I somehow never could find time to have.

Ellen

When I joined Group, I had been an assistant professor at UCB for a little under two years. I was not only the first woman to join the department but also, at that time, the only faculty member who was not a full professor. Christine Guthrie and Keith Yamamoto invited me to become part of a group of men and women, all involved in the life sciences, who met twice a month to discuss professional problems. It was an offer I couldn't refuse.

I went to my first Group meeting because I was flattered at the invitation and intrigued with the idea of having a source of support and advice, which some part of me knew I needed but which had eluded me. I stayed in Group because of the depth and variety of issues discussed, the wonderful and varied personalities and experiences, and the much needed reminder and assurance that I was neither alone nor crazy. As time passed, I came to believe that I had something to offer the others—an important element of the group process for each of us. The personal relationships have deepened, while the wisdom and value of the professional advice have persisted. Humor is a consistent highlight in Group—for example, the concept of a "To Don't List," for someone trying to learn not always to say yes when given a new assignment, and "Premorse," as the catchword for regretting something that hasn't happened yet. We frequently end up laughing the most deeply after discussions of the most difficult issues and situations.

My anecdote is that at the end of 1981 I learned that my department was going to deny me tenure. This event may demonstrate that Group is not omnipotent with regard to academic achievement and success, but the more important message is that I came through this truly devastating experience with more self-confidence and self-respect than I had had before it happened. Group process was crucial to my being able to evaluate honestly what had happened and why, what my role in it was, and what I could not control. With Group's help I was able to determine where I next wanted to go professionally, and to go there with pride and imagination.

Helen

When I joined Group in October 1980 at the invitation of Ellen Daniell, it was my first experience with a professional problem-solving group. There were six woman members at that time, and most of the issues were related to their professions. Some of the areas were pragmatic—how to move beyond writing blocks, how to recruit, interview, and hire a lab technician, how to help troubled graduate students, how to deal with faculty politics, how to find effective ways to handle university bureaucracy, and the like. Sharing our experiences and using the format of the Radical Psychiatry movement has enabled us to learn from one another and to become more empowered as individuals.[2] As we spent more time together—for me it's now been fourteen years—the problem solving has moved into the personal area as well. Together we have faced the death of loved ones, failed relationships, marriage woes, parenting concerns, illness, and aging. Today Group's work is intermingled, the professional with the personal. I marvel at what we have created together—a safe, caring, thoughtful, and intelligent place in which to be our true selves, a place where we can air and solve our problems with compassionate hearts. And I'll tell you what I cherish most about Group—the laughter and the genuine friendship around the table at the end of each session. For no matter how hard or agonizing or difficult the work has been that evening, we always end up with laughter and with a great appreciation for one another.

Suzanne

At my twenty-fifth college reunion, the keynote speaker, Betty Friedan, referred to women of my age group as the cusp generation. We were the ones caught between changing images of a woman's role, halfway between *Ozzie and Harriet* and *Murphy Brown*. We were the generation that could still remember wearing girdles. Certainly I was caught in the middle of this revolution in women's aspirations. While I was in graduate school, I met and married a brilliant astrophysicist and the love of my life. As his career progressed, I followed him around the country, raising three children and working part-time as a kind of superannuated postdoc. At age forty I realized that I would be soon out of a job in science. What to do? I began taking courses at night in the vague hope of becoming an accountant.

Then I had two lucky breaks. First, I was invited to join a small nonprofit medical research institute where all the scientists worked in my specialty, the study of human vision. Smith Kettlewell is a very supportive environment for a beginning principal investigator, particularly if she is forty years old and doesn't have a clue about how to run a laboratory. It is, however, a soft-money place, which means you have to obtain grants to buy equipment, to purchase technical support, and to pay your own salary. Living on soft money is a little bit like coming up for tenure every two years in perpetuity. To join Smith Kettlewell seemed to require more courage than I had at the time. Which brings me to my second lucky break. I was invited to join Group. These women, all struggling with their own insecurities, talked me into taking a chance. What did I have to lose? If I didn't get funded, I could always become an accountant next year. So I joined Smith Kettlewell and began applying for grants, and every year for the past fourteen years I have managed to postpone becoming an accountant for one more year. In fact, I have been amazingly successful in science. What I learned from this group is that it is better to take the chance of doing what you want than to accept second best.

Mimi

I'm Dr. "Premorse." Science is an aesthetic experience for me, and what drives me is awe for the beauty of natural form and a tremendous curiosity about how it works. That aesthetic appreciation led me to be an art major when I was an undergraduate, but the curiosity led me to switch to science. I study the fluid mechanics and engineering of organisms. For me it's exciting to explore natural form in a quantitative way. There is a price to pay, however, for using unconventional interdisciplinary approaches, and that is, you are always an outsider. A practical consequence is that you don't fit job descriptions at the back of *Science* magazine, and you fall between the cracks at funding agencies. But my passion for natural form kept me doing research when I was unemployed. When I finally got a job at Berkeley, I was funding my research with my paycheck.

I joined Group as a frightened little assistant professor, terrified that I wouldn't get tenure. Group provided me with practical advice and moral support that I needed to carry on submitting grant proposals over and over and over and over again, until I finally got my foot in the funding door. But Group is not just a nurturing squad of cheerleaders. This is one tough bunch of ladies, and they do not tolerate whining. We see right through one another's bullshit. Whenever I start moaning about the latest roadblock to my ability to do research or to teach well, Group reminds me that there is no "them" out there trying to get me. When I find myself grousing about my latest frustrations with work, I often say, "What would Group say about this one?" Group is also great for providing reality checks. When my department put me up for tenure early, I was convinced they were doing it to get rid of me early, but Group beat some sense into me. And when I was convinced that my telephone call from the MacArthur Foundation was a practical joke, they simply ignored my fussing and drank a toast.

Judith

I first became interested in science while in high school, as the result of two excellent teachers, one in chemistry and one in physics. When I bemoan having to teach yet another organic chemistry class to five hundred premeds, I remember how important teaching is. I was in a special class for academically talented students. We had all our classes together and were socially insulated from the rest of the school. Within this group there was a lot of competition but also a great deal of camaraderie. Thinking back, that first experience of a support group gave me the courage and self-confidence to go forward and test myself in college. The feature that stands out the most about my group of high school peers was that we seriously challenged one another while remaining close friends and intimates.

When I went to Berkeley twenty years later, it was as the first woman in the physical sciences. Talk about feeling isolated! At the time, I was also a single parent of two preadolescent sons, yet another source of social isolation. I had met Chris Guthrie through a common friend, but they did not believe I would join Group, so the possibility was not raised for several years. I had convinced myself, and apparently others, that I didn't need any support—you know, the superwoman syndrome. This position came at tremendous cost, to both me and my family. I have now been in Group for about thirteen years and in the process have learned repeatedly to rely on the collective wisdom and caring of the members to navigate never-ending personal and professional crises—they do seem to go on and on, I must say.

During the past five to seven years at Berkeley, I've had a fair measure of success in my career and only now feel that I have reached my prime professionally. What is remarkable to me is the degree to which I have experienced a feminization of my perspective, learning to trust the voices of other women and myself. I know there is a strong correlation between our trust of our inner voice and our capacity to be our most creative and productive in science. Group has shown me the possibilities that come out of this feminine voice through hard-nosed feedback dished out in an environment that is both supportive and loving. Because most of us have been in Group for a long time, we have become a family of sorts. If there is a danger, I think it's that we may stop being honest in order to protect one another. Fortunately this hasn't happened so far.

Now, even at the prime of my career, I still encounter obstacles in my work environment and have to fight constantly to avoid feeling marginalized. Only last week I was denied a promotion that I thought would be routine and would have placed me on a par with my male colleagues at the same level. In the midst of my distress, I called a member of Group who was able to give me the balance and empowerment I needed to make some decisions. Increasingly, I feel I am proceeding with the collective knowledge and wisdom of the women in Group. Its pull is so strong that it has at times kept me from exploring job opportunities outside the Bay Area. In fact, that may be the most serious drawback of a Group such as ours. Once established, it's very hard to leave.

Carol

Being the newest member of the group, I really don't have a lot of Group history to share with you, but I do get to have the last word. So I'd like to tell you how I came to join and what it's meant to me, and to suggest reasons why even those of you who feel you are doing very well might like to join or start a group like this.

I was at the University of Wisconsin (Madison) for twenty years and was acknowledged as a leader both on campus and in the scientific community. If anyone had suggested to me that I might want to join a group like this, I would have thanked them politely, but privately I would have thought that being in such a group was at best a waste of time and at worst crazy.

I moved to the Bay Area a year ago, primarily because of my long-term relationship with Hatch Echols, a professor of molecular biology at Berkeley. Our happiness when I got a faculty position at UCSF was shattered when we found out that Hatch had lung cancer. I took care of him in Berkeley for the last few months of his life and then started my new job.

The circumstances of my move were very different from what I had anticipated, but luckily I had agreed to join Group as a way of forming new relationships. These women, a few of whom were mere acquaintances and some of whom were total strangers, proved to be my lifeline. They provided me with professional support, explaining how to make my voice heard in this new environment, where all of a sudden I had no status. At the same time they helped me grieve my loss and take steps toward recovery. They gave me books to read and left flowers on my doorstep so that I would know I was not alone. The first time I missed Group because of my busy teaching schedule, I was thrilled I felt I could survive for two extra weeks without their input. That was a real measure of how far I'd come.

Now, a year later, I'm much more secure, I know how the university works, and sometimes I'm even happy. The idea of leaving Group, however, is the furthest thing from my mind. I now know that I am incredibly lucky to belong and that my previous reluctance to join such a group was based simply on ignorance of its advantages.

There are several reasons I recommend Group's process to everyone. First, because we are encouraged to work or report on something at each meeting, I take stock of the good and bad in my personal and professional life at least every other week. This allows me to pinpoint problems before they become too big to be solved and to identify things that make me happy that might, under other circumstances, pass unnoticed. Second, my fellow Group members provide not only constructive, honest, and critical feedback about how to approach my concerns but also the emotional and practical support I need to carry out my decisions. As a consequence I face personal and professional problems more clearly. Group empowers me to change when necessary to make a better life for myself. It is somewhat ironic that because of the tragedy of losing Hatch, he left me his most valuable legacy—his friends in Group.

These are the stories that we chose to explain Group to others. The audience and sponsors were enthusiastic and the presentation sparked lots of questions. The evolution of Group before the time of these talks illustrates that its flexibility lent it strength. Group membership turned over almost completely in the first three years after its founding. Although individuals certainly change the character of the group, our basic objectives have remained the same, and it was these objectives that attracted new members to replace departing ones.

Of course members are a key factor in deciding whether to form a new group or join an existing one. When Keith and Christine described the group to me, we were all assistant professors in the Bay Area. I admired them both. I had been introduced to Christine when I was a postdoc and she was relatively new to the faculty. Having heard of her and her work in a research field close to my own, I was thrilled to meet her, and a bit intimidated. She was my ideal of the scientist I wanted to be—brilliant, fearless, and unapologetically womanly.

About a year after I went to Berkeley, I heard that Christine had been hospitalized at Christmastime following a suicide attempt. The grapevine was unclear on details, but the core message was that the pressures of trying to succeed in a high-powered and cutthroat environment had led to a breakdown. I was sad and shocked, but not mystified. I remember feeling that I understood her action in light of my own crazed existence. Being really sick sometimes seemed the only way to slow down. If I felt surprise, it was not that an assistant professor had reached such a desperate place but that it was Christine, who in my eyes had it all together. The idea of Group became all the more compelling to me, knowing that it was part of the survival plan of a brilliant, high-profile person whom I had thought to be completely in charge of her life.

I was attracted to Group because I felt lonely and unsupported at work and because I liked the stated purpose of helping members succeed in their professions. Group seemed to share my belief that there was more to life than work, yet it focused on survival in professional life rather than escape from it. I had been reluctant to involve myself in escape activities, for fear they might claim too much of my time. I could justify investing in a group that would help me do my job better as well as relieve my anxieties.

I went to my first meeting with a mixture of excitement and trepidation. Everyone seemed to be frank and open in presenting problems and giving advice to others. I learned that the structure of meetings assured everyone a voice. The political philosophy of Radical Psychiatry, the movement that had inspired Group, was a mystery to me, but I appreciated the existence of procedural rules to keep us focused and feeling safe. All the members were embroiled in the whirlwind of academic politics and competition. People at different career stages were concerned about performing their jobs and creating a supportive work environment. This was a reassuring contrast to my one previous group experience, where my professional status isolated me. In those meetings, women who were struggling to pay the rent expressed resentment of my comparatively well-paid job and doubted that I needed their support. At the beginning of Group, I feared being an outsider because I worked at a different institution from the rest, but I rapidly became comfortable talking about my professional worries. Being from a different campus may even have made it easier for me. The others could provide meaningful advice because my problems were familiar, but the specifics weren't intertwined with everyone else's issues.

Members left, some because they moved from the area, others because they didn't find Group process useful or they had met the challenge that brought them. As membership fluctuated, we determined that seven or eight was the most workable size. Although I was sad to see people leave, I did not worry that Group might dissolve until Keith Yamamoto left in early 1980, followed less than six months later by another founding member, Toni King. Keith announced his departure shortly after being awarded tenure. I inferred that, in light of his achievement, he felt Group was no longer useful to him, and my reaction was resentment. I was also distressed that it would no longer be a mixed-gender group. But rather than dissolving, we invited new members. Group has grown as a branching chain, new members often known only to one person before joining. Fresh attitudes and ideas have enriched the mix.

I want to emphasize that two striking aspects of our organization—close friendships and longevity—are not essential elements of a successful group. It is evident from this book that the current members of Group are intimate friends. But we were not close friends when we started, and Group was useful and productive before these friendships developed. People who remained in Group for only a few years also benefited. Keith, who was in the audience at our 1994 presentation, addressed this point in the discussion that followed: "There were some people who entered the group with specific sorts of work that they wanted to do . . . and then when they did it they finished that and they left. There were others obviously who were invested in this in different ways and were involved in the long term. What they gained from it continued to evolve and I think that's part of the profundity of what we've heard tonight. There were a few people who came into the group and discovered that this was not something that was really helping them or that they really identified with and they left. I think the secret . . . is that while you see a lot of stability in front of you there has also been a lot of dynamics. . . . People who have passed through . . . would feel that they gained as much from Group as the people you just heard speak."

3 *Facing Disaster*

This autobiographical chapter illustrates how Group functions in times of individual turmoil, in this instance the termination of my faculty position at the University of California, Berkeley, and the resulting need to reevaluate my life goals. I would surely have made different choices during this crisis had I not been in Group at the time. This was not because others directed my actions but because I was able to work through many possibilities and consider alternatives with their protection and critical feedback. I ended up with a sense of choice and self-direction that contrasted sharply with the lost and helpless feeling of professional defeat.

I begin this tale early, with how I came to be a scientist, because science brought me to Group and also to the crisis that

Group helped me survive. I did not grow up with a sense of being called to research. There was no science tradition in my immediate family, and my first serious professional goal was to be an actress. Now in my fifties, I am learning to describe myself as a writer. In the years between, I chose to pursue science, a pursuit that claimed me and shaped my life. For eighteen years I defined myself as a scientist, and for thirteen more I held jobs that depended on my scientific knowledge and training. My husband and many other close friends are scientists. "Scientist" is part of my self-image.

My love of science had its earliest roots in high school, although I didn't acknowledge the connection until later. A fabulous ninth-grade teacher awakened me to biology through innovative lectures and assigned reading. I enjoyed chemistry and math, and although I was scared of physics I could handle it. But I thought of myself as an "English/history" person, spending hours in libraries writing long term papers, not entering science fairs like my "math/science" friends. I didn't even learn to use a slide rule (a must in those days before hand-held calculators) until college. Consistent with the view that my academic strengths lay in the humanities, I arrived at college (Swarthmore) intending to be a political science major. Two weeks into freshman year, I replaced one social science course with introductory biology, a move I described as "keeping my options open," because the idea of majoring in science had by then occurred to me.

By the end of the year the option had solidified into a decision to major in chemistry, a rigorous background that would prepare me for biomedical research.

Biology and chemistry labs satisfied my desire to produce something. I found myself impatient with studying humanities and social science, because they demanded too much analysis and commentary on the original work of others. I took great pleasure in the long afternoons in laboratories, although they were hard work. I was drawn to the puzzle aspect of science, the act of going through a set of procedures and getting a result. Reading my college diaries now, I am inclined to laugh at the young woman who exulted in freshman lab courses, but I remember my awe and fascination. Science captured both my heart and mind. Later in college and at graduate school I enjoyed designing experiments and working late at night to complete a preparation, record data, or take samples. Science was an adventure, and it was exciting to be involved.

The people in science were a major attraction. I identified with a group that studied in the chemistry, physics, and math library. We sat at favorite tables, raced to the dining hall at the last minute when an experiment ran long, and discussed equations and reaction mechanisms in the hallway during study breaks. The junior and senior honors chemistry majors encouraged me to follow in their footsteps. I rapidly came to appreciate chemistry for its own sake, loving especially the rigor and mathematical clarity of theoretical chemistry.

I was the only woman in most of my chemistry and physics seminars. I was as well prepared scholastically as the men, although I felt disadvantaged in physics labs because I had never done electrical wiring or built things in pursuit of a hobby. Although I didn't see being a lone woman as a particular plus to my chosen path, neither did I view it as a negative. A story from a letter home illustrates this. I baked brownies for an all-day lab session, and an enthusiastic classmate asked whether I'd be willing to do so again. I agreed, and for the rest of the semester, I furnished baked goods. Every week someone would bring me a cake or cookie mix for the next week. I loved baking, and the arrangement increased my sense of camaraderie with the guys. My lack of concern that this was a "girl" thing to do was typical of my experience of Swarthmore, which had a long tradition of gender equality in academic pursuits.

Having chosen science, I found I excelled at it, despite constant doubts and fears of failure. My roommate Jean said that before every exam I feared failure, after every exam I was sure I had failed, and when the grades came out I would get mostly As. My professors encouraged me to go to graduate school. Most of my friends were going to schools in the East, but I applied principally to programs in California, which had the multiple attractions of distance, climate, and reports of student activism. Looking back after thirty years, most of them lived in California, I see it as a choice of adventure over a path I perceived as more predictable. I ultimately chose the University of California San Diego (UCSD) over Berkeley and Stanford, because the chemistry department there included opportunities in biology. I wasn't sure where my interests lay.

My years at UCSD, in La Jolla, were a wonderful experience. The first year I did short research projects with six professors. When it came time to choose an adviser, I decided to work with John Abelson, a smart, dynamic assistant professor whose lab was busy and full of fun as well as excitement about the research. My thesis research involved a virus called Mu, which infects *E. coli* bacteria. The underlying purpose of the work was to uncover the mechanisms by which organisms control the expression of their genes. The project led me into several collaborations with scientists in other labs and institutions, and I began to see myself as part of a community of interesting people doing important research. This sense of belonging deepened as I began to attend scientific meetings to discuss my work. My life felt balanced and full. Although I was often in the lab past midnight, I took midday breaks to run or swim, went on long bicycle rides, and cooked dinners with friends on weekends. Despite a few crises of confidence, I took for granted that when I attained my degree I would find a postdoctoral fellowship and then a university faculty position.

After four years, I was close to completing my PhD, and it was time for the next steps. As was the case when I was selecting a graduate school, I again sought adventure. Scientifically, I decided to move from the bacterial virus work with which I was familiar to the study of animal cells and viruses. These systems were less well developed and more difficult to work with, but they were intriguing models for human cancer, and I was eager to do something different and challenging. Personally, I wanted to find a lab in Europe. From among several possibilities, I chose to work with a scientist who was at the Cold Spring Harbor Laboratories, on Long Island, but who would soon be moving to a cancer research facility outside London. I was to spend only a few months on Long Island before going to England with the project. During my interview at the Cold Spring facilities, I found the lab scientifically enthralling and physically beautiful, although a little stifling socially. But I figured it would be bearable, even fun, for a short time. After I arrived, my new boss reneged on his plan to move, and I spent my total fellowship on Long Island.

A second event at that time was even more critical to my future course than the choice of a postdoc. Before I left the West Coast, I was invited by the Molecular Biology Department at UC Berkeley to give a talk on my graduate work and be interviewed for an open faculty position there. I had no intention of looking for a university job until I had a few years of postdoctoral work under my belt, but my thesis adviser told me it would be good experience to do the interview. He said I should feel no pressure, "because that department will never hire a woman." I had the interview on my way to New York. My seminar got high praise, the visit went well, and I headed east excited about the future. I received Berkeley's job offer the next May, five months into my postdoctoral stint. Getting a job based on my graduate research was a coup!

Cold Spring Harbor was a big change from university life. It was a small establishment, and everyone lived on the grounds or very nearby. People talked about science all the time or gossiped about the private life of others. My adviser, who worked long hours, expected to see everyone in his group back in the lab in the evenings. One colleague was ridiculed because he babysat for his young daughter one or two nights a week so his wife could pursue her career in the performing arts. The directors fostered competition between "The Lab" and groups at other places doing similar work. I had seen competition before, but this felt nastier. When someone returned from a conference and reported that investigators at the National Institutes of Health were on the wrong track about some phenomenon our scientists had figured out, team members seemed to take more pleasure from the competitors' error than from their own achievement. Competition among people within the institution was also nurtured. It was not unusual for two postdoctoral fellows to discover that they were working on the same scientific problem, one or both having been assured that it was solely his or her project to develop and pursue. The heads of different units vied for recognition, derided one another, and some forbade their research fellows to talk to those in the other labs. In this environment, I developed a new set of feelings about science. I was beset by doubts about my ability and was no longer certain that my devotion to science was sufficient to achieve the success I craved and had anticipated.

Countering the negatives were spikes of elation over my science when it was going well. I was still thrilled by new experimental results, and enjoyed perfecting manipulations at the lab bench. My colleagues were smart and dedicated, and many seemed to thrive there. I made a few close friendships. But more and more I was happiest being alone in the lab, avoiding the paranoia I developed when others were around. This attitude prevented me from taking full advantage of the stimulating exchange of ideas that was the most powerful aspect of Cold Spring Harbor. I continued to work feverishly and began to give talks about my new research. Cold Spring Harbor is a conference center, and I enjoyed the continuous flow of scientists who came through for meetings and seminars. I flipped between excitement and despair in a kind of manic way, asking in my journal, "Am I nearing some sort of breakdown or a breakthrough? And how will I know which it is?"

After briefly considering another job possibility, I accepted Berkeley's offer and agreed to cut my postdoctoral work to a total of two years in order to satisfy their need for a new faculty member. Once this decision was made, I spent time planning the details of a new laboratory and writing a grant request to support my future research. From the point of view of professional development, I would have been better served by a third postdoctoral year, which would have allowed me to learn more and publish a couple of papers in my new field. A renowned scientist whom I greatly respected advised me to delay taking a faculty job, but I was too eager to leave Cold Spring Harbor, too eager to return to California, and too triumphant about Berkeley's offer to seriously reconsider my options. In retrospect, it seems that I felt impelled to move forward as fast as possible. I had gotten a PhD, I wanted to do research, and I took exciting opportunities. What I did not do was to consider carefully which course would be best in the long run for building my career in science.

And so I moved to Berkeley. Happy as I was to be making a change, this was "out of the frying pan and into the fire." Now I had to run a lab, attract students to my lab and advise them, teach, and deal with university politics. Although the scientific competition with colleagues was less vicious because there was a larger community and a greater variety of research under way, the pressure to perform was greater, and I was even more alone.

I had gotten some clues that I might not be totally at home in my new department. On my initial interview, several faculty members had been obsessed or uncomfortable with my being a woman. The chair was very proud that the department had hired a woman on his watch and boasted of having done so before Berkeley's rules on gender-based affirmative action went into effect. I was not only the first woman on the department's faculty but also the only member who wasn't a full professor, the most recent previous junior faculty member having left after being denied tenure.

From the start, I was exposed to departmental politics and was shocked by the open disdain that certain faculty members showed toward one another. I didn't know the history that led to the bad interactions, and it was very odd to be treated to "confidential" recitals of various colleagues' failings and misdeeds by others trying to garner my sympathy for their position. When I arrived, half the faculty members were lobbying the dean to relieve the chair of his duties and appoint a replacement, a campaign that succeeded the following year. In another case, an emeritus professor was writing a stream of vituperative letters accusing the department of treating him unfairly by reducing his lab space. I tried either to laugh at or ignore the situation; it seemed not just vicious but downright crazy.

My major task that first winter was to deliver six lectures for the graduate course on molecular biology of eukaryotic organisms (animals and plants). This course was slated to be my future contribution to departmental teaching; the next year I would be responsible for all the course planning and lectures. The chair and other faculty had told me that the current course was poorly designed and asked me to teach subjects of more fundamental interest, exposing students to key research topics in molecular and cellular biology. I had strong opinions about what was important, but I had much new material to master, because I was fairly new to the study of animal cell systems. My commitment and my confidence that I could improve the course didn't overcome my stage fright. I was overwhelmed by the tasks of giving lectures and preparing challenging exam questions. My first student evaluations praised my selection of subject matter but were critical of my presentation. Several comments described my lectures as somewhat confusing; this was perhaps a reflection of my state of mind.

I got off to a better start running a lab. I had no direct experience, but I knew what I needed to do and assumed I could handle it. My own graduate adviser, who had been a new assistant professor when I started to work with him, was a role model. The department at Berkeley had a fabulous support staff, from facilities and storeroom managers who helped me set up the lab to the office administrators who led me through the grant application process. I had negotiated for the university to provide certain equipment that was waiting when I arrived, and campus granting sources paid for supplies until my first big grant from the National Institutes of Health was funded in June. I was a bit daunted by the prospect of competing against scientists I had worked with as a postdoc, but I was truly excited to be starting work on my own.

As soon as the grant money was available, I hired a technician. We had been recommended to each other, and it was clear in the interview that we would be friends. A senior faculty member from another department warned me that employing a friend could be a mistake, but I rejected his advice. I wanted my lab to be a home, and my students and employees to be my friends. This was politically satisfying, and I enjoyed the camaraderie of the lab, but I did later come to appreciate the difficulties inherent in this choice. Being a protective friend was more natural to me than being a taskmaster and boss, and it was often hard to be clear, even in my own mind, what I needed from staff and students professionally if I was involved in their lives personally.

I worked hard and took on many tasks, serving on various departmental committees. When the department created an undergraduate major, I became one of two faculty advisers to the program. I created experiments for a new undergraduate laboratory course that was offered for the first time in 1979; students asked me to give out diplomas at the first undergraduate graduation from the department. I served on graduate student exam committees in other departments as well as my own and felt both honored and overburdened because so many students requested me as examiner.

I had a life outside work, too. I fell in love the spring of my arrival in Berkeley, with someone who was neither in science nor at the university. He was also not really available, and I ended up hurt, lonely, and very sad by the first of July. For the next two years, I shut down the idea of developing a long-term attachment. I was too busy learning to be a professor to cope with a serious relationship, and I ended up using the pain as a protective shell.

As a diversion, I went dancing, which I had enjoyed as a graduate student. Folk dancing always proved a way to meet interesting people. I joined in the rental of a ski cabin in the mountains. I moved from my solitary apartment to share a house with another woman, an art historian. Although we kept separate schedules, we would occasionally eat or spend an evening together; I enjoyed her companionship.

Time passed. Two more assistant professors joined the department, so I was no longer as isolated by my age and rank. One of the two was an exceptionally talented woman, and I took pride in having paved the way for the hiring of a second woman. (A few years before, one of my colleagues had told a talented female postdoc that she should not bother to apply for the opening after my being hired because "we already have a woman.") I became interested in a new line of research within the animal virus field and directed successful grant applications to that endeavor. More graduate students joined my lab group and one or more undergraduates came each year to perform research for a senior thesis. I loved working with the students in the lab, both directing and learning from them. My classroom teaching improved. I poured my heart into making the core graduate course relevant and interesting and for several years shared the responsibility with another assistant professor. My efforts were rewarded: student attendance increased, evaluations became positive, and I derived satisfaction from figuring out how to present the material. I continued to feel alienated from the department and appalled by faculty politics, but I survived.

Two landmark changes in my life fostered my survival. In 1977, I joined Group. My work there focused on my doubts about science as the career for me. I got valuable practical advice on teaching, advising students, and departmental politics and renewed my determination to succeed as a faculty member, where getting tenure was the next big hurdle. In December 1980, I married David Gelfand, with whom I had been living for two years. David, a biochemist who had trained at UCSD and UCSF, was director of recombinant molecular biology at Cetus, a new biotechnology company. The level of my anxiety about research and teaching was much reduced by having Group's support on specific issues and a partner who was committed to science and confident in me.

At noon on December 7, 1981, my professional life changed dramatically. The chair informed me that the department had voted not to recommend me for tenure. I was caught off guard, although of course I knew I was up for tenure. I had received materials on the review process, including the university "Fairness Safety Guide." I had provided materials and submitted names of scientists from whom the chair should solicit outside letters evaluating my research. But I wasn't aware of the precise timeline on which the decision would be made. And I was distracted, having spent most of November in bed with hepatitis A. I had returned to work half days only the previous week, crawling back to bed in the afternoons. I was quite pleased with myself that particular morning because I'd completed revisions that ensured publication of a significant manuscript in the *Proceedings of the National Academy of Sciences*. So when I was called to see the chair in his office, I did not immediately anticipate the reason.

He began, "I have some bad news I have to tell you," and I instantly knew. As he elaborated, my brain seemed to split in two, one half trying to listen and figure out how to reply, the other fantasizing about how I could shock him out of his smooth professional posture. How would he react if I shouted "What a relief!" or "Thank God I have an excuse to escape this place!"? I was not relieved; I felt I was in a nightmare and imagined saying something to disrupt the story. I don't remember what I actually did say, however. Eleven tenured faculty members who could seldom agree on anything had concluded, though not quite unanimously, that I was not worthy of being promoted to their august ranks. Six years after I had joined the department, a committee had recommended that I should not stay, and the majority had agreed.

The chair was very proper and was probably trying to be kind.

He handed me a copy of the committee's written report, explaining that I had the right to make written comments within a few days and certain rights to protest the decision. I had the impression that he considered protest undignified and assumed I would not do such a thing. I tried to read through the report, but it was too hard. I opted for getting out of the office before crying.

Helen Wittmer—friend, confidante, and Group member—was waiting for me when I emerged from the chair's office. As administrative assistant to him, she was the only person other than the tenured faculty who knew what had transpired. Helen was my first line of support and defense against utter despair. I couldn't have had better company to listen to my resentful and bewildered reactions and to deliver much needed understanding and compassion. In heavy fog we took a walk up the hill behind campus on paths I had never taken. Never having returned to them, I've always had a sense that the trails we walked that day existed only in an alternate universe.

Although I'd been afraid I would get teary in the chair's office, I was unable to cry for several days. I felt numb and kept focusing on practical issues because feelings of inadequacy were too frightening to let in. What would I do about the ten weeks of teaching that I was to start in January? How could I present my committee report on courses and curriculum to the faculty later that week? I had just learned that my whole career had crashed and burned, yet it was these details that occupied my mind. I left Helen, drove to Cetus to tell David the bad news, and then went to a doctor's appointment and home for a nap.

I desperately needed Group and got through the afternoon sustained by the knowledge of a meeting scheduled that very evening. This was presumably a coincidence, though I wondered whether Helen somehow influenced the timing of my meeting with the chair so that I would have immediate organized feedback. (She always said she had not.) I knew that talking with Group would give me the best chance of figuring out how to deal with this disaster.

My initial reactions centered on the written committee report, which I read over and over as the afternoon dragged on. Each time I read it I felt insulted in some new way (even twenty years later, when I dug it out of a file to write this chapter, I still felt misused). There was a simple conclusion: "Having given due consideration to all the available evidence, we recommend against the promotion of Assistant Professor Ellen Daniell to tenure, since her lack of standing as a productive research scientist is not sufficiently compensated by what we judge to be her adequate performance in teaching and her excellent to outstanding University service." This formal, civilized statement summarized a discount of six years' hard work and accomplishment. I felt that the committee neither understood nor valued what I had done. I focused on the many factual inaccuracies but absorbed the condescension as well. They appeared to have thrown in positive statements about my enthusiasm for science, my outstanding university service, and the "hospitable atmosphere" of my lab to give me something to feel good about in the face of their disdain for my research. Every conclusion had a negative slant. They said my research was "sound," that I chose to "work on challenging and interesting problems," but that I was "not well focused on the goal." I felt like a little kid reading the critical demeanor report that counselors at summer camp had sent to my parents, wanting to protest, "But it wasn't *like* that!"

The teaching evaluation offended me most. In the winter before my tenure committee convened, only fifteen students had registered for my course. The committee held me responsible for the low turnout, although they acknowledged that our own department's first-year graduate class (obligatory attendees) held only six students and that "there was a conflict in scheduling with an advanced biochemistry course taught by a distinguished Visiting Professor." Although written student evaluations of my teaching had become predominantly positive over the years, the committee seemed to have gone out of their way to get negative comments. "Because of the difficulty of forming a realistic evaluation of an instructor's performance entirely on the basis of the responses entered in course rating questionnaires, we discretely [*sic*] sought the personal opinions of several graduate students." No faculty member had attended one of my lectures to develop his own opinion, which I would have welcomed, and which I understood to be standard practice in many departments.

I also agonized over the research evaluation. I would have found their negative review more palatable if the case had relied on the outside letters. Instead, they made their own comparison of my studies to other work in the field, giving two examples. One was only tangentially related to my interests, and the other was incorrectly portrayed as illustrating my failure to make a conclusion that in fact I had both made and published. I had been advised three years earlier that I needed to strengthen my résumé with publications, especially on a new topic I had started to study. I had done that, publishing several papers in the new area, one of them an innovative bridge between two fields of inquiry. None of it appeared to have helped. It seemed they wanted me to be a different scientist from the one I had been all along. Describing the work of others, they asserted, "We would not, of course, criticize Daniell for not herself having made these particular discoveries," but I fumed. That was exactly what they were doing! My departure would give them a slot to hire someone else.

At last evening came, and I could go to Group. Helen and Mimi went ten miles out of their way to pick me up so I wouldn't have to drive. When we started the meeting, I asked for forty minutes (indicating a crisis), and I spoke last, knowing that it would be hard for anyone else to work after my announcement.

I started with the facts and tried to move on to how I felt, as Group often encouraged me to do, because my tendency was to get carried away with details and ignore my emotions. But this evening I was struggling with numbness and scared of what would happen when I started to feel anything. It was safer to be stoical. Group advised that I not worry about when I was going to run the gamut of emotions, to just let it happen in its own time. "Experience your feelings as they come, when they come, and cherish them because they are yours, even if painful."

I felt I couldn't grasp what had happened. It wasn't that I thought I could alter reality, more that I didn't yet believe it. Group's advice, to accept what had happened and start from there, seemed close to impossible, but it was the right approach. I read the committee report to them, remaining calm when enumerating their research criticisms; accepting criticism was part of doing research. But I became very emotional when I got to the teaching section. I was infuriated that it cited as evidence of the low quality of my teaching that students working with one particular faculty member in another department did not choose to take my course. These students, the report said, were studying molecular aspects of animal virology, "a subject to which much time is devoted in [my course]." In fact, I devoted very little time to that subject because I believed that other topics were more important, and even a cursory review of my course prospectus would have told them that. With Group I could vent my resentment, proposing that a more accurate statement would be "members of the faculty who are virologists are disappointed that this is not a virology course that their friends with narrow interests will encourage their students to take."

I was delighted to hear Group's indignant reactions to aspects of the report, but even more valuable to me that night was their empathy for my wounded pride and sense of betrayal. Notwithstanding our commitment to directed work and practical problem solving, we know that sometimes an outpouring of support is warranted, and that night everyone knew I needed to vent and get sympathy before getting down to strategy.

Group identified my instinct to take on the committee summary as a truth about myself, my inclination to believe that "I am not being recommended for tenure, therefore I'm no good." Their feedback included affirmations to help me combat this tendency to believe the worst. Mimi said, "You are so many wonderful things, have so many facets, are so warm and caring. Your science is just a little piece of that." Judith stated the direct contradiction to the conclusion that I was a failure: "The rules of the institution can *not* be the rules you judge yourself by." Christine offered, "There are many people, not just Group, who love you, think highly of you, *respect* you as a person and a scientist." That statement helped me to let in the calls and gestures of support from other scientific friends over the next few weeks, to embrace, instead of avoiding, their condolences.

I sought Group's advice on getting through the next days and weeks. I was very concerned about telling my students the news and how to guide one student through completion of his thesis. I also needed to prepare a written response to the committee report, and above all I didn't want to lose my focus on restoring my health. "You should stay home for the rest of the week," everyone (but me) concurred. It was a Tuesday, and staying out for three days felt like a dereliction of duty. Group pushed me to agree not to go back to my office the next day. Answering their questions about why I wanted to go in, I realized that it was very important to me to tell the news to the other two untenured professors and to my students before they heard from anyone else. Group pointed out that I could do that from home, and I made it the first order of business the next day.

The other critical task I wanted to work on with Group was my formal response to the committee report. In the safety of our meetings, I could voice all my suspicions about which faculty member had directed the review and who was responsible for the most egregious errors. Most of this would be neither appropriate nor effective in arguing my case, but anger and despair were the emotions I could best access, and giving them free rein made it easier to think things through. Group advised, "Write from the heart, include all thoughts no matter how inappropriate, then clean it up." This is often solid advice on writing, and it worked in this case. A second ground rule was that I should not men-tion names. Doing that would antagonize people, and inaccurate guesses would weaken my statements. Summing up the approach I wanted to take, someone said, "Dignity, grace, and integrity are the watchwords for getting through." A tall order, but I believed I could do it with Group's help.

The next day I followed the plan we had discussed. I phoned Liz Blackburn, the second woman faculty member in the department, at home. I was very fond of Liz, and she was familiar with my work and publications. She reacted with obvious shock, "Well, what do they want?" When I told her not to worry, she retorted, "I think very highly of you, therefore I shall indeed worry." She was enraged at our tenured colleagues and pulled no punches as she talked about it. Helen later told me that Liz had talked at length to her about how angry she was and how concerned for me. I arranged to meet my most senior graduate student, Marty Fedor, off campus. Marty was furious that the review featured negative student comments, yet no one had questioned her about the training she had received from me. She also thought, with more philosophy than I could muster, that I would have "many interesting options after Berkeley"; she didn't seem to think my professional life was over.

Staying away from work while taking time to inform significant people of my "disgrace" was a good strategy, but it was a relief to spend that first afternoon with Helen, who took the afternoon off to lunch and walk with me. With a member of Group I could express my worst thoughts and expect them to be heard but not taken as signs that I was falling apart. Helen was happy to see me. "I was afraid you would disappear like a wisp of smoke, and you didn't."

I prepared a list of the things that I saw as demonstrably unfair. The report had ignored positive written student evaluations of recent years and hadn't mentioned the positive reviews of my most recent grant application, which had received a high rating and been fully funded. The grant review concluded, "Daniell has a good publication record, is positively cited and appears well on the way to becoming the authority on adenovirus chromatin." These omissions from the report had left me very suspicious about the rest of the process. I met with the chair, who passed on my comments to the dean in a memo.

Group was my lifeline. At our next meeting, I reported my progress, including my first uncomfortable encounters with faculty members. One had refused to look at me when we met in the photocopy room, leaving quickly with a mumbled hello. Another had said, "I think you will be happier with life out of the university," which I took as an attempt to assuage his guilt. When I reported my polite responses, Group cautioned me against being "too nice." "That's okay as long as you don't get sucked into it. It's fine to protect yourself, but don't get into trying to make things easy for everyone else." As usual, Group humor cheered me in a tough situation. We imagined my saying to one senior colleague whose excessive solicitude for my mental well-being irritated me, "How are *you?* I've been concerned that you might be depressed since you are so close to retirement." Although I never actually said it, I got comfort from our laughter.

Exhausted, I worked with Group on how to keep going and how to make the best use of my limited energy. "I wish I could just do nothing until I get tired of doing nothing." My graduate student who was in the middle of his thesis research had asked whether he needed to find another lab in which to work. Teaching my upcoming course seemed a monumental burden. I didn't want to deal with any of the tasks that confronted me: research, teaching, or correspondence—none of it. Judith advised that I define a few goals for the rest of my stay at Berkeley. I followed this tack through the next several meetings.

I felt deeply humiliated and embarrassed with respect to people in and out of my department. Group was invaluable at helping me identify which actions might be useful and which were meaningless or likely to make me feel worse. I was frantic to know whether the department members at large had read my whole file or just the committee summary. I envisioned that people had only the summary information and that they believed it. I was paralyzed by the thought that had I been a faculty member reading the report, I too would have voted against granting me tenure. I thought everyone must be looking at me and thinking what a loser I was. The only way to escape would be to leave science and cut off contact with scientists. I even struggled with feelings of shame and envy with regard to the successes of other Group members. When Mimi reported that she'd gotten a major grant funded and Judith talked about her efforts to develop a greater sense of service to her department and university, I was sad that I had no more chances to do these things and felt ineffective in light of their continued achievements. These feelings were difficult to confess, but when I did, Group came through, seeing my reactions as natural and letting me work through them without resenting or misunderstanding me.

My illness actually helped me get through the first two months. I was so focused on getting well that I had an uncharacteristic reserve of attention to my own well-being, the unusual perspective that my health was more important than my job. Being sick also led to some comic relief and external perspective. The very day that I got the news, I had kept a doctor's appointment. To his "How are you today?" I replied, "Pretty awful. I've just learned I'm not going to get tenure at Berkeley." The doctor, a man about my age, said, "But that doesn't affect your *job* does it?" I explained that, indeed, it meant I had lost my job, or more precisely that I would lose it in eighteen months. Here was an educated professional who wasn't familiar with the "up or out" principle of tenure. Maybe there were others! It made a good story for Group, and in the ensuing months I often pictured the puzzled doctor whose reaction belied my conviction that the whole world was looking at me with pity.

I entered a kind of professional limbo as a result of the odd characteristic of a negative tenure decision that I could keep my job for more than a year after effectively being fired. (This seems even stranger to me now that I've worked in a corporate environment where a departing employee may be escorted to the door within minutes of dismissal to prevent sabotage or theft of company information.) Walking across campus I felt rejected by the physical space of the university, not just the faculty and administration. Suzanne, who had worked on campus for many years in a non-tenure-track position after completing her doctorate, helped me identify my reaction. "If you are not faculty, you don't feel like a real person on campus, because people don't treat you as one." A similar sense of invalidation has motivated many people to leave academia for industry.

I had three paths to follow, which were not entirely compatible. First, I wanted to continue to do my job with a degree of professionalism until I left. Second, I had to begin to pull away from the university and look forward to what I might do next. Third, there was still the possibility of launching a formal protest and asking a university committee to reverse the department's decision. There were puzzles associated with each of the three. Right after Christmas I had to start teaching, which felt hard after being told I did it so poorly. I was proud of the course, however, and regardless of my feelings about the department, I retained a sense of obligation to the students. Group supported my dedication to this task because it was personally satisfying to me but suggested that I conserve my energy by asking other faculty to give some of the lectures. My spirits were boosted by high enrollment, and my teaching evaluations from students at the end of the course were my best ever, many of them glowingly enthusiastic. Several cited incoherent guest lectures as the only major problem with the course. I took spiteful pleasure in this, since the guests were senior faculty members speaking about their areas of expertise. Once the course was over, Group urged me to let go of the sense of ownership that kept me worrying about its future. I needed to be done with it.

As I continued to brood on what I could have done differently to attain tenure, Group advised me to look beyond the university, to "find a new trajectory." Trying to consider alternatives, I came to understand that I had internalized the message that being on a university faculty was the only worthwhile career. It wasn't so much that I thought I was incapable of doing anything else, but that no other endeavor was as valuable. The elitist lessons of academia built on my natural sense of obligation to repay the world for my good fortune. There were, as well, parts of university life that I was reluctant to give up, first among them being interactions with students. While encouraging me to think broadly about who I could be, Group reminded me that I did not have to plan the specifics immediately. I reported my irritation at people making unsolicited suggestions about the wonderful possibilities now open to me. Mimi used a memorable analogy to reinforce Group's advice to ignore them all. "If your husband had died, people wouldn't be suggesting great singles bars."

The question of whether to protest the decision at a higher level was hard. My initial inclination had been to accept it without further action, hoping to send the message that I wanted nothing more to do with the department. I felt battered and sensed that protesting would subject me to more pain. Also, Group had helped me recognize that I had made real choices along the way that had a bearing on the outcome. I had done research the way I wanted to do it, not according to someone else's ideas. Perhaps I could honor those choices by just letting the decision stand.

On the other hand, David, some members of Group, and other friends urged me to challenge what they saw as unfair treatment by the department. Perhaps by protesting I would feel that I'd "done all I could" and be able to put behind me the mistakes I kept realizing I'd made. For example, I had not understood the need to ensure that the scientists writing letters of recommendation were aware of all the aspects of my research and the progress my lab had made. I should have sent each of them preprints of papers that were not yet included in my file. I had not solicited the advice of my departmental colleagues. I had not even used Group to the best possible advantage. They could have helped me by reviewing my tenure materials, but I had assumed naïvely that I had the situation well in hand.

Deciding reluctantly to launch a formal protest, I initiated the process according to university procedures, asking the chair to request additional letters. I selected scientists familiar with my newer work and sent each one a description of my research objectives and other supporting materials. I consulted with Group about the tone of my cover letters, wanting to communicate facts without being apologetic or embarrassed. Filing the protest compounded my emotional difficulties with day-to-day life in the department. I felt I didn't want to see or speak to any of the faculty ever again, but it seemed foolish to burn my bridges in case the tenure decision was reversed. I maintained a kind of uneasy coexistence, feeling variously angry, sad, incompetent, and superior. I was also lonely.

Protesting was terribly difficult, and it is still difficult to write about. Professionals are supposed to be tough, take their lumps, and move on. Nonetheless, the process was useful, even though the decision was not altered, because it helped me access my feelings and acknowledge my conviction that I had been treated badly. Although I may have bolstered my self-respect by setting the record straight, I do not believe there was a real possibility of reversal. On July 16, 1982, I received a formal letter from the

Berkeley chancellor with the final decision that the university would not undertake a reconsideration of the department's decision. I felt sad and terrified, and this time I cried often. But the finality was better than the uncertainty.

From that point, I had one more year on the faculty. I made a contract to "take care of commitments, leave time for myself." Gradually, I disentangled myself from the rage and was able to think more about "what next?" than about "what they did to me." The open-ended aspect of choosing a future was a bit overwhelming. "I have a chance to be happy if only I can make the right choice instead of the wrong one." Group suggested ways to approach the situation from a fresh viewpoint. "Claim this period as a unique, magic time in which to make real choices." "The future doesn't have to look a certain way; you can do anything." To take the weight off the decision-making aspect of my thinking, Group suggested that I think of myself as "accumulating possibilities." "You are sampling, getting information. Nothing is final." Christine visualized it as cruising, "being out of gas and coasting."

I felt a profound lack of personal power, which made fulfilling commitments harder. The once-routine task of writing a paper, submitting it to a journal for critical review, and then resubmitting it after responding to reviewers' comments was now a struggle. I told Group, "I have no firm underpinnings from which to deal with any kind of rejection." The obligation to help my students complete their experiments and publish the results helped me maintain my balance, and I did continue to write and publish.

Group reminded me that I had found it hard to enjoy science in the years before the tenure decision, because there was so much riding on it. Judith suggested that now that survival was no longer the operative goal, I could find out if I enjoyed science more with less pressure. These considerations led to a short-term plan with long-term options. I had for several years been interested in switching fields to work on the molecular biology of plants, so I arranged to spend a year in the lab of a colleague in the genetics department who worked on maize. The National Science Foundation supported this plan by agreeing to redirect funds originally granted for my virus work to support the maize project. I took my first and only sabbatical leave, spending three months in a plant research lab at Yale. This reawakened my pure excitement about being engaged in scientific research, and I felt as if I'd escaped from a poisoned atmosphere. When I returned to Berkeley for my final three months as professor, all the bad feelings came back, and I doubted I could ever achieve anything in science again. I told Group, "I've been walking around being terribly brave, and suddenly I'm exhausted." I cleared out my office and moved to my new lab home across campus. My research in the maize lab was exhilarating, and I made some interesting discoveries, but a year later I left Berkeley and research science for good to take an administrative job in a biotechnology company.

I quit research and academia because my experience at Berkeley had left me in doubt about my ability to be a productive scientist. I could no longer summon the energy and dedication to launch a research program, frame the important questions, and design experiments to answer them. I had begun at Cold Spring Harbor to feel ambivalent about whether I would be happy with a life in research, but I had not then doubted my capacity to be a scientist if I chose to be. After Berkeley, I felt totally disenchanted with the endeavor. Since leaving research science, I have enjoyed my work more, felt more accomplished, and received much more external approval than I ever did as a professor. It took me ten or fifteen years to fully understand that my success and comfort elsewhere didn't necessarily mean that I was not a good scientist and teacher. Although the focus of my work moved on to other arenas, Group, with its knowledge of all that had gone before, was able to help me reach that understanding.

Only drastic changes could have produced a different result for me. I was thrust into a situation for which I had no training, and for which none was available. (Our group was not set up to take on professional training, although I think that could be a legitimate and useful focus for a group.) There was little or no recognition at that time that new faculty could benefit from having a dedicated mentor to identify pitfalls. I needed an advocate within the department, someone to take responsibility for my case, advise me on my progress, and counter my detractors. Neither the former chair nor any of the others who had so enthusiastically persuaded me to go to Berkeley took steps to be mentor or advocate, and I was too naïve to recognize the lack. Group's support and advice made it possible for me to get past devastation and move on. My experience not only sensitized us to the importance of mentoring but also taught the lesson that each of us is responsible for learning the rules of the institution(s) in which we work and for knowing how to take care of ourselves and act in our own behalf.

Part 2 Group Work

4 *Accepting . . . Liking . . . Celebrating*

APPRECIATING OURSELVES AND BEING
ENTITLED TO SUCCESS

- *Grant yourself authority to do whatever you are doing.*
- *I am entitled to be myself. I'm entitled to be successful.*

Among all the topics we address, those that have most engaged me may be the importance of asking for what one wants, taking credit for achievements, and feeling entitled to success. Learning to accept and forgive one's own missteps is only the beginning. We encourage each other to move beyond self-acceptance to self-appreciation and from there to celebration of our accomplishments. Celebration requires a sense of personal entitlement, a belief that we deserve to be happy, to succeed, and to have what we need to survive. It is hard to rejoice in achieving something if we don't believe we are entitled to it.

Group's focus on self-celebration and entitlement arose not from abstract discussions of attaining personal happiness but rather from practical work on professional survival. To do a job well, people must be able to ask for all that they want and need, without apology. That stance requires a sense of entitlement we often lack. Mimi told us that when she thinks of making a request, she hears her mother reproving, "Who do you think you are, the Queen of Sheba?" This is the childhood lesson of not asking for too much lest we seem greedy. We suggested the affirmation "Maybe I *am* the Queen of Sheba" to prepare for making a request (or a demand). This has become a Group catchphrase. Whatever we need—a salary increase, equipment, secretarial help, or a more convenient travel schedule—we have to be unequivocal in our belief in the validity of the request. We know there will be plenty of negative energy from other sources; we shouldn't contribute to it. A contract on this subject could include the sentence "I will act as if what *I* want is most important."

I have plenty of experience with the basic work of forgiving and accepting mistakes because of my apparently boundless capacity for chastising myself, especially over losing things. Group helped me cut off the blame in one episode that concludes with one of my favorite strokes (a positive observation; see Chapter 16). While looking for a job at the end of the Berkeley years, I laid my briefcase on top of my car and drove off. I never found it. The case was my husband's first gift to me, and the contents in-

cluded important work and my current Group notebook. I used
this event to dwell on a host of weaknesses. I reviewed my obsession with writing things down, my fear of forgetting anything I
didn't have in writing, and my shame at being thoughtless, careless, not valuing David's gift enough, and so on. Judith pegged
my loss as misplaced anxiety. "It gives you something bad to focus
on other than your fears around your job search." Suzanne characterized it as "an opportunity to tell yourself you are not competent, at a time when you need to feel especially competent, a
kind of self-sabotage." These insights, which put the situation in
the context of a stressful time, helped me feel less crazy and think
of forgiving myself. When it came time for strokes, someone
said, "It was brilliant and brave of you to throw away your briefcase, getting rid of that old stuff as you head out in a new direction." I gaped, then exploded in laughter. Group recycles this
stroke frequently, adapting it to other situations. The ultimate
way to deal with an unforgiving pig (self-criticism; see Chapter
16) is to celebrate the criticized action. We don't give up on trying to alter behavior where indicated—I really would prefer not
to lose things—we just try not to beat ourselves up about it.

Compassion is key to accepting and liking oneself. One night,
reporting extreme overload at work, I said, "I clearly need to be
more organized and I would like help with that." I thought this
was a good resolve, but instead of applauding, Christine challenged me: "Ellen, you are good at identifying the problem, but

not so good at doing something about it. You set things up so that the solution is not fixing the problem, but being a better person. So there is no relationship between taking some action and making the situation better, because the only solution is *being* better." This way to look at my efforts at self-improvement made sense. My insights about being overloaded were accurate, but I had turned them into a judgment that I was a disorganized person. Others weighed in: "You never turn your compassion toward yourself. You declare that being out of control is your major fault; then there is no point in either getting more rest or making more plans because you are not sufficiently organized yet." My approach of having to "be better" took away my power to do anything to change the immediate situation. "So what do I do?" I demanded, a little irked because others seemed to see the issues so clearly while I was confused. Christine answered, "First, do nothing, just observe." I didn't need to figure out the reason that I was so hard on myself. I should just notice and be compassionate. As I ended my work exhausted, I said, "I don't know how it would feel to treat myself with compassion, but this clearly isn't it!" I made a contract to "choose to listen to and identify my feelings. Don't act, or analyze, just hear them." I titled a page in my notebook "Ways to feel okay about being Ellen."

A practical aspect of self-celebration is taking credit for what one has done, both day by day and over long periods. Most Group members have made progress at recognizing our own achievements, which we have found to be as difficult as getting public recognition. When I am exhausted and overwhelmed by things I haven't done, I usually remember to consider the things I *have* gotten done. Sometimes it's a matter of noticing the good news. When Mimi was overburdened with a backlog of manuscripts that needed her attention, she rephrased her analysis after talking to Group. "The reason there is a crunch of papers is that I did all this neat research, *not* that I am irresponsible and haven't written the papers."

It is not unusual for women to avoid taking credit for their successes. A friend to whom I talked about my writing cautioned me not to ascribe my life to luck or accident. She referred me to Jill Ker Conway's book *When Memory Speaks*, which describes how remarkable women such as Jane Addams and Florence Nightingale consciously played down the intention that underlay their achievements. Their private diaries record prodigious planning and determination, but their memoirs written for the public eye suggest that they stumbled into their careers. These women apparently chose to characterize their achievements as lucky accident as a shield from the attack they surely expected for having set unfeminine goals. Conway's treatise provides a historical context for Group's focus on each taking credit for what she does.[1]

Our work has revealed several pitfalls that block us from taking credit appropriately. First, we do many things that don't appear to be part of our formal job description. A day might be spent helping a colleague with a paper, ordering equipment, going to a committee meeting that appears a waste of time, and comforting a student whose experiment (or relationship) has failed. This is all hard work, and probably essential, and much of it may further career goals. Yet each of us has felt that she accomplished nothing because she spent no time on critical planned tasks. So nothing is ticked off the "To Do" list. As an aid, Group has introduced the concept of not waiting until something is completed to count it as an accomplishment. Suzanne described this as taking "a modest reward for pieces that are done." This technique helps with getting started on a task, breaking it up into "chewable bits" as we put it, so that the smaller tasks are less daunting. It is also a fundamental part of creating "do-able contracts" (see Chapter 16). With practice, it gets easier to remember that most things we spend time on should count as accomplishments.

We also recognize a tendency to leave off a list of achievements those things that were easy or that we didn't specifically list as goals. I will say, "Oh that! I didn't do anything, *that* just happened," yet I never omit items from my "Needs Improvement" list. Judith illustrated how sneaky the mind can be in this discounting process when she told us, only half in jest, "Guggenheims [prestigious fellowships for faculty members] couldn't be as big a deal as I thought because I got one."

Finally, in taking stock of what we've done, we have had to learn not to discount interruptions. Time spent giving advice or assistance, often requested without prior scheduling, is part of most jobs. As some of us took on administrative responsibilities, we learned that the need to acknowledge interruptions gets more and more important in management. To the group contract "Write down the interruptions and cross them off," I added the addendum "Give myself extra points for the especially long and frustrating ones."

Even in Group meetings where our awareness should be especially sharp, problems may take precedence over an accomplishment. Helen worked on financial worries one evening and tossed off as an afterthought that she'd received a professional achievement award and was proud and pleased. When Mimi had received a job offer that she had been hoping for and talking about for months, Group heard the news as a "by the way" at the end of her work on another topic. Judith suggested we need to "get off on the good news" as individuals and as a group. "Much of the work we do is about feeling bad despite the good things that happen." Strokes at the end of our meetings allow us to highlight one another's good news and recognize or reiterate accomplishments.

Although self-recognition of accomplishments comes first, we must also publicly take credit where credit is due; indeed, this is a professional responsibility with important consequences. In most pursuits, advancement and job satisfaction are affected by the image we present to others. Failure to let others know what we've done encourages a perception that we haven't done much. Being self-effacing in our public persona has an enormous cost, and it is no surprise that women raised in the fifties and sixties find this a problem. Many of us were taught as little girls that "it isn't nice to brag," which, as we grow up, becomes reluctance to say "I'm doing well." We are afraid that assuming success will lay us open to ridicule. It's not modest. This may be exacerbated by the support and comfort that come our way when we're not doing so well, giving the insidious message that we'll get those emotional goodies *only* when we're in a nonachieving place. Being strong looks less attractive if it means we will forgo attentive concern. We need to believe that we will be honored for our achievements *and* that we must acknowledge and celebrate them even if appreciation is not forthcoming. In summary: "Act as if there were no question but that you would succeed. There is nothing to lose by assuming success, and there is everything to gain. Assuming success empowers."

When Suzanne's independent research was taking off and getting recognition, she raised concerns about presenting herself as successful. "I've previously been gentle and generous. Now I need to be impressive, but I find that it's accompanied by a sneering mode. Is it necessary when you're becoming a success to cut other people into little pieces?" The effort to own power while maintaining valued elements of personal style such as gentleness and generosity is a central Group issue. The goal is to acknowledge what we have achieved and be honestly proud, without losing ourselves.

At some level, I linked my own reluctance to present myself as smart and powerful as uniquely my problem, a result of feeling weak as a scientist. It was an eye-opener when Judith worked on difficulties preparing to give a talk on a major scientific breakthrough in her lab. Worried about the feelings of the scientists whose hypothesis she was about to disprove, she had drafted a presentation that was tentative rather than forceful. "I'm going to challenge the dogma, so I'm scared that in an effort to avoid angering people I'll present too weak a picture." She resolved to give her subject the tone it deserved. She reported afterward that her revised talk was a smashing success, but she "felt bad, really sad for the people who got it wrong. They were humiliated at this meeting." Group's response to these feelings was that her concern for others was appropriate and valuable, but she must not let it overcome the reality of her own success.

So we women see that we avoid taking credit and acting successful because we want to be liked and be "nice" and because we enjoy having others take care of us. Another impediment to taking credit and acting successful may be a fundamental feeling that we are not entitled to be *too* happy. When someone said, "Everything is going well in the lab, and I feel terrible," Christine's feedback was that maybe the bad feelings arose because she had surpassed expected limits. "We are trained to think we can only have so much success and are disconcerted when things go well on multiple fronts." The key is to believe that we are entitled to be successful. If we don't, then our accomplishments will never be satisfying, because we will feel undeserving instead of exultant. It is amazing how hard this is. Group has a very clear role here, to bring someone back on track when she loses touch with her sense of entitlement.

I close this chapter with an incident that shows how easily we can slip up and find an excuse to hide achievements. In 2001, Group gave a panel discussion to women faculty at the University of California San Francisco. We discussed not mentioning all our awards and honors in the introductions lest it seem we were boasting in front of our audience of equally competent, high-achieving women. Then, we realized the folly of that attitude and laughed at ourselves. We would be ignoring the commitment to honor our accomplishments. Furthermore, we knew that some women shy away from support groups because they perceive that to need advice and support is somehow an indicator of weakness or failure, so the professional status of individual Group members was relevant to our message. We believe in the power of co-operation, and many of us are confident that our participation in Group has played a significant role in our successes. So we listed the honors, without apology, but shared our irresolution with the audience.

We need to believe that we are entitled to be successful and to couple that belief with self-confidence and acceptance in order to take risks, fail, and try again. We want to choose goals that satisfy us, enjoy the process of reaching those goals, and decide when to respond to the demands of others. An affirmation sums this up: "We have permission to be whatever we are."

5 *A Serious Mind and a Light Heart*

RESPECTING INSTINCT AND PERSONAL GOALS

- *The need to please is never satisfied. Please yourself.*
- *I feel like an emotional cafeteria, responding to what people want of me.*

The urge to please others at the expense of what matters to us is a constant impediment to honoring our own style and instincts. The impulse to look excessively to others for direction and approval affects our professional choices and our behavior with friends and family. In my experience, the desire to please sneaks up, so that I find I am working from an image (often inaccurate) of what others want. The work described here is closely tied to issues about making choices and celebrating who we are. The last part of this chapter looks at how respecting one's own wishes is critical to maintaining personal creativity in a professional context.

One Group member who had directly confronted this issue at work brought it to my attention for the first time. A university department secretary with many people claiming her time, she decided to close her office door for two hours each day. She needed uninterrupted time to work on long-term projects that were her responsibility, but some of the faculty and staff were irritated by her limited accessibility even for those few hours. She told us that when she was "behind the nice mask," she gave everyone else's opinion greater weight than her own. Now she was taking it off, exerting her right to decide what tasks were most important, even in the context of an administrative support position. Pursuing this theme, she talked about the difficulty of being a secretary, finding that respect for her work was not usually forthcoming from others. "I want external approval, but I want not to need it."

That last statement struck me as apt for anyone struggling to let her own intentions and wisdom direct her actions. It is, of course, important to have our accomplishments appreciated by others, as a practical part of getting funded, promoted, and published. Doing well in the world is about external approval. But if applause from others becomes the goal, we can no longer define success as we see it. Suzanne found this work applicable to her own current struggles as a first-time principal investigator running a lab and research program. She reported that she wanted "the respect of my peers more than I want *self* respect. I don't know how good my science is; I only know what others think of it. This makes it very difficult to take criticism." The feedback was that it is self-defeating to work toward a goal of being a scientist with the respect of her peers. "The goal must say what you want to be, not how you want to be looked at." Continuing this work in a future meeting, Suzanne said that she didn't know who she was because she tended to be defined by others. She wanted to find out. Christine suggested that "Who am I?" was too cosmic a question and that "What do I want?" would be more workable. Words of encouragement also held a note of caution about expectations; "when you've been pleasing other people all your life, it takes time to find what you want."

These discussions were timely for me in the context of having failed to please the university and my department faculty after six years of trying. Almost everyone could relate an experience of trying to please people, doing things for *them*, instead of toward an end or in one's own interest. "Pleasing others is a false setup. What you need to do is grant your own significance, accept that *you* are important, and can direct a life." Following this advice has demanded my continued vigilance. When I decided to work in an administrative position in a biotechnology company, my former faculty colleagues greeted the news with raised eyebrows and even sneers. Although they had denied me the research career I wanted in academia, they disapproved of my alternative. Seeing how absurd it was to let those people judge my choices reinforced my intention to choose a new direction simply because it intrigued me.

After my first year at Cetus I felt unappreciated, partly because people didn't care much about the personnel function, so it seemed that no matter how well I did, it wouldn't be important. I had been asked to draft my own performance appraisal and was beset by the sense that I would be judged on criteria that didn't reflect my most important achievements. Group suggested that I do two appraisals—"one to discuss with your boss, the other just for you which acknowledges everything good you've done." I developed a contract to value myself and my work regardless of the appreciation of others. "Visualize yourself as you would ideally like to be, and then project that image. Let go of your concern about what 'they' think of you. Don't look for feedback and acceptance from outside, because your very need for it blocks it. Do the job well for itself." I was subsequently rewarded by a promotion to personnel director despite my lack of formal training for the job. Several years later, still sensitive to external judgments, I summarized a difficult management struggle and declared my feelings with slightly desperate humor. "What I am and how I do it is not enough. They're right, I'm wrong, and I'm on *their* side." Group's advice in this situation—advice that is quite frequently applicable—was to be very clear with myself about what I wanted and thought was right, so that if I were to act differently to satisfy someone else, I would do so consciously.

It is easy to slip into a pattern of response. External demands are sufficient to take up all the available time, and then some. Responding to what others want carries much less risk of rejection than does stepping out with our own ideas, especially if those ideas lead us to neglect someone else's demands. But in responding to the needs of others, we neglect our own and may end up, for example, as Suzanne did when she spent days helping a colleague with a grant request although her own grant deadline was a month earlier and her need for the funds greater.

Mimi recognized the drain of constantly being aware of what others want. "I have been failing to notice the present. I want to create moments to remember but I don't notice my life because I'm not in it. I'm a little machine doing the things on my list." In contrast, when performing field research at a remote site she remembered every detail, and her chronic headaches and neck aches ceased as well. She determined that the difference was that nobody made demands on her while she was doing fieldwork because she couldn't be reached by anyone at the university. Her dilemma was to live with the demands and to coexist with others while maintaining a personal sense of what is important. Of course there are roadblocks, as Mimi reported soon after; "I'm trying to do what's important to me, but "they" keep having other ideas.

I have an image of making boundaries and putting inside them only those things I value. Then other people dump extra stuff inside the boundaries."

Symbolic acts can be meaningful. I applauded when Helen announced that she had given up answering the phone "May I help you?" and Mimi had changed the message on her answering machine to express less eagerness that the caller leave a message. They were both moving away from the assumption that to receive a phone call is to serve at the will of the caller. Whenever I reach Mimi's message (which is not unpleasant, just matter-of-fact, with no promise of response), I remind myself that I too can protect myself from miscellaneous requests. My challenge now as a writer is to not always answer the telephone when it rings, just because I'm at home.

A major barrier to finding out what you really want is the possibility that the knowledge will disrupt your life. In talking to Group, I once realized my fear that, if I really honored my desires, I might not do any of the things I was doing; yet my life was comfortable, and a major change would be hard. I also recognized that when I thought about appreciating myself and doing what pleased me, my pigs told me I was being self-absorbed. Group pressed me to describe what I associated with that accusation. I responded "egotism, being a selfish person." Christine said, "You see it as selfish, rather than the only sane way to be."

The pursuit of creative research illustrates the importance of pleasing ourselves. Those of us practicing science know it is important to follow scientific passion and intuition, respect one's own ideas and personal style, and have the courage to try new directions. Along the line, however, we've been told it is unscientific to be intuitive, so as women we neglect intuition, which may be a particular strength. Intuition is a component of good science and of many other creative endeavors. Deciding on the next experiment (or business project) is often not obvious; it may require a mental leap. Design and execution of a proper experiment and interpretation of results can be taught. The ability to choose a new direction that is at the heart of creativity is much harder to teach.[1]

The university and the scientific granting system often make being true to your own direction difficult. Scientists are encouraged to pursue experiments that they know will work because they need to show productivity to get the next grant funded, and those pursuits may lead away from the most creative science. Judith made a big impression on me soon after she joined Group when she spoke of her determination to stick with her own conclusions scientifically. "I want to take risks and allow myself to make mistakes, while staying in touch with reality." She had written a paper that her colleagues criticized for having concluded too much from her data. "But I'm not willing to play by the rules and give everyone what they want. I'm going to submit the paper as I see it." Judith concluded, with a sigh, that risks take energy: "It's hard to keep up your beliefs for which you are being criticized." She submitted the paper, it was accepted for publication, and her interpretations ultimately turned out to be correct, and provable. Other Group members have affirmed in different ways their determination to do science out of curiosity and delight, not for status and productivity. There is an important relationship between pleasing oneself and self-acceptance: "You have to love yourself enough to have it be okay when you flop. Then you don't have to do only the safe thing."

Mimi and Suzanne worked on the importance of sticking to a personal style of research in how they chose to report on their results. They each expressed discomfort with the fact that they wrote fewer scientific papers than was the norm among their successful colleagues, though each believed she had chosen the most appropriate method of writing up her research. By talking this through together, they established that they wanted to write "long, elegant papers that are easy to understand." They developed a shared affirmation: "The way we have chosen to do our science is the right strategy for the science we've chosen to do."

We have shared experiences and fears about exploring new research directions on unfamiliar topics. When I began to study plants, I was terrified of being trivial in a new field, but learning about this new topic made me more excited about science than I had been in years. Judith, planning a sabbatical, resolved, "I want to learn something totally new, or learn new techniques to apply to my current problems." Someone else described doing science as "feeling that when you get into the right place a unique vision will come. If you can tolerate things being out of control and allow yourself to think, it will happen." The topic of uneasy feelings that arise when working on unfamiliar problems sparked a very animated discussion about exploration of new fields. "Discomfort and lack of understanding are part of the creative process. It's important not to have preconceived notions about science, and in a new field you are more likely to be able to avoid them." And citing the need to acknowledge accomplishment: "Usually the most interesting things in science are at the interface of fields. Give the lab and yourself credit for doing something hard."

Work that Christine did on honoring personal style in the professional world provided a guide to me in working and writing. She was planning a sabbatical, a period of freedom from the usual responsibilities of teaching and university business. She told us she wanted to use the time to get more in touch with her child side than with her analytical side, to recapture an approach to science with a sense of wonder. In short, to make the sabbatical hers instead of trying to figure out what she *should* do. Her contract to "have a serious mind and a light heart" expressed her desire to accomplish much, to look at science in an open way, and to enjoy her work. This is the spirit with which I would like to approach professional endeavors, indeed all aspects of my life.

6 Off Balance and Out of Control

MANAGING TIME AND ESTABLISHING EQUILIBRIUM

- *Enjoyment of almost any activity is linked to having enough time in which to do it.*
- *Life is not to be lived as if it were an emergency.*
- *I'm already going much too fast. I won't go any faster. I am going to slow down.*

I came to Group one night tired and stressed, asking, "Why do I move so fast, keep working all the time?" One explanation was "to avoid the stuff you don't want to think hard about." Another was "habit—being busy is what we do. If we aren't working, we must be depressed." For me and for others in Group busyness often seems a way of ignoring deep dissatisfactions, of avoiding the need to solve big problems in life. That night I made a contract: "I will make sure I have some amount of time with *nothing* to do." This resolve has never been easy for me to maintain, but in framing it I accepted the premise that stopping to breathe was worthwhile, a valuable goal.

The interwoven issues of time, busyness, and control are a major focus of Group work. Our meetings set up a context of respect for everyone's time by requiring that each person designate how many minutes she intends to work and then reminding her when that time is up. Our discussions of time and busyness have taken several directions. We have tried to determine the roots of the need to be busy. We have developed the idea of "maintaining equilibrium" instead of "managing time" as we look for solutions. Practical solutions include giving any project the amount of time needed to complete it; giving complete attention to the task at hand; and learning to say no. More and more, we've recognized that a major reason to pull back from obsessive busyness is to allow ourselves the time to consider what we want, to contemplate the choices open to us.

The experience of life as overcrowded is widespread, as demonstrated by the cascade of books, courses, seminars, and counseling services dedicated to time management. Our time is filled with activities and responsibilities and our minds with ideas, information, and lists of things to do. Women who have struggled for expanded opportunity have a hard time admitting that they are busier than they want to be. Rejoicing in having a wider scope of activities available to us than did women thirty years ago, we find it complicated to say "but I want to relax, find the joy in my life." Not expecting that life will get simpler, or that there will be fewer things to do, we need to learn to handle the demands on us. Group's perspective has been invaluable in my search for the right balance. Colleagues and family, although supportive, have a stake in the completion of tasks that affect them. Group focuses solely on my welfare and can recognize when it is more important for me to breathe and contemplate than to get something (or everything) accomplished.

Judith's work on busyness illustrates how the need to fill all spare moments develops and persists in our lives. When her younger son left home and she had an empty nest after many years of being a single mother, she reported, "I'm working harder and harder." She resolved "to think about what I want to do, to choose. I want to be clear, not driven." Despite this contract, she continued for several years to have the sense that she was upping the ante for success, becoming more and more active and ambitious, a workaholic. Although she enjoyed the excitement and success, she sensed that she didn't want to live her life so narrowly. Returning from a retreat one summer, she described her deep commitment to expanding the personal side of her life. Her plan was clear . . . until the fall semester started. She had to write a huge article, prepare a talk for a conference, teach a new course to five hundred students, and write a status report on her major grant. The rapidity with which she put personal issues on hold gave her pause. She said, "It's frightening to see how I attempt to avoid myself. My greatest comfort is in knowing the next few months have no free time." Because the tasks were important and unavoidable, it made no sense to turn away from work just then. Group advice, which she made into a contract, was "Maybe it is time just to listen, pay attention to how it feels, *experience* being so busy, without trying to change right now." When the crunch was over, Judith returned to pursuing her personal commitment.

Mimi has brought many poignant examples of how unforeseen events constantly impact our efforts to maintain equilibrium in professional life. She was told scant days before a lab course was about to begin that the number of teaching assistants assigned to it had been reduced, which increased her personal workload. Another time, an administrative assistant deleted an important prerequisite in the catalogue listing for her lecture course without consulting her. Fifty students signed up, came to the first lecture, and were angry because they were unprepared and had to drop out, messing up the schedules they had designed to accommodate the course. At one Group meeting she described her professional life as a game of tennis. "While I'm trying to play, little gnomes are trying to untie my shoes. It doesn't matter if I hit the ball or not because I'm so busy dealing with the gnomes." These incidents were incredibly frustrating for Mimi, but telling us about them and infusing some humor into the recital made them somewhat more bearable.

Group advised that the only thing to do was to stop trying to fix matters. "Under these conditions you can't teach as you would like to. Anger bears too high a cost, so you have to not try so hard to do it right." Group has persistently advised Mimi to take permission to do things less well so that she can survive. Compromising excellence is tough to put into practice, because it feels inherently unsatisfying. Several of us work on the difficulty of letting things go as the only solution to overburdened schedules. We identified and named an "Out of Control Pig," whose message is, "I am doing too much, and therefore doing it all poorly. Everyone else with this job can do it fine." (We remind one another that we don't know what internal hysteria those other, seemingly calm, accomplished, and clearly in-control people are experiencing.) In time, we recognized that getting everything under control implies a certain endless struggle, a battle that we were unlikely to win. After yet another administrative snafu that threatened to soak up Mimi's time, we helped her devise an affirmation and contract: "It's not my fault and I can't fix it. I will do the things I think are most important today given the resources I have here and now."

"You can do something for an hour and enjoy it, or you can do something until you finish it and enjoy it, but you can't finish it in an hour and enjoy it." We often revisit this practical affirmation, which emphasizes that how we deal with time affects our level of satisfaction with what we do. Setting out to finish a task in a prescribed amount of time is an invitation to failure and encourages the doer to watch the passage of time instead of relishing the process. We all agreed that we rarely experienced the luxury of doing one thing until it is done. We take notice of the exceptional times—when someone enjoys the freedom to work in the lab during a sabbatical, or focuses solely on one specific critical task, or gives herself up to a book or activity on a vacation. There is a contract that I think strikes at the core of this issue: "Commit yourself to a project without flogging yourself about how long it takes. Let the project define the time, once you've decided to do it. Make peace with how long it takes, *and* with the result, whatever it is. If you don't commit yourself to the endeavor, get out of it. When there's not a choice [e.g., because of an inescapable deadline], commit to doing it, then enjoy what there is enjoyable in it." One of the rewards is that we are generally most creative working at whatever rate feels comfortable.

"I have the right to do one thing at a time. I have the right to be doing what I'm doing." This affirmation addresses the difficulty of committing to an activity by mentally putting aside the previous and next things on the list. When I was personnel director, I reported that I feared making mistakes because I moved too fast from one issue or problem to another. Group advised, "Take a breath as you are about to declare something finished as well as when you are starting something new. Separate events from one another. Let go of the next thing until you start it. Time is what keeps everything from happening at once—so *use it*!" I came up with a mental ritual for putting away conversations and projects so that I could give full attention to the matter at hand. I thought of a double click of the computer mouse to close and save the image. My rule was not to have more than one person or problem on my virtual desktop at once.

Prioritizing and planning are necessary elements of equilibrium. Because getting to activities we've designated as low priority is usually impossible, we have to couple prioritization with commitment, so that we set limits, which has been a recurring theme in Group work. The *I Ching* says that limits are necessary for fostering creativity, because "without limitations you are overwhelmed by possibilities." If we don't let go of some things, we make ourselves miserable by always regretting things not done. When someone working on establishing and accepting limits expressed the intention of doing "fewer things better," Christine suggested the affirmation "Less is more." Mimi quickly added the caveat "Less is more but none is not enough"—lest any of us be tempted to a life of sloth.

Setting limits and making them stick requires learning to say no with confidence and comfort. We have tried to do this by recognizing our right to choose what we do and by making conscious choices rather than doing things by default. When we choose to do something, we are empowered to do it. Saying no is also a right and should be equally empowering. One reminder I found particularly important is that "it is okay to say no to legitimate requests." I hadn't thought of it that way before and recognized my previous need to convince myself that a request was unreasonable if I refused it. Christine developed another guideline for saying no when she reported that she had decided not to participate in a book project that interested her greatly, because she needed time for other things in her life. "Say no in the context of the alternative, to make room for something else."

Lists prove useful when setting limits, and we often encourage one another to make them. In my most high-pressure work experiences, I used lists to allay anxieties, writing everything down, then singling out the most important things and letting go of the others. Nonetheless, when a week of interviewing job candidates loomed, I worried about how much time that would take and was oppressed by the thought of letting people down by not finishing other tasks. Group suggested that I figure out how long the interviewing would take, "then eliminate that number of hours (or a few more, just in case) of other work and tell the relevant people that certain things will not get done the next week." I had not actually accepted the fact that I had to eliminate something. We talked about making lists of things we wouldn't do, which became a favorite device, the "To Don't List."

Inevitably, there are stumbling blocks in using lists to prioritize and limit. One person reported that the hard thing about making a list is that it made her admit she wasn't going to do something. "If it all seems important to you, it's a downer to make that admission." And another said that she had made a list of all the things she wanted to work on in the coming fall. "I'm so disorganized I lost the list and can't remember what was on it." I found myself completely dependent on lists and complained, "I can't do anything anymore that I don't write down."

If we don't actively prioritize, we find we let outside demands determine how we spend our time. Mimi described this as prioritizing by guilt. "I'm out there with the bucket and shovel putting out brushfires, saying, 'I'm ready world, do it to me,' instead of making choices." We tend to deal only with the loudest demands and squeakiest wheels. The recognition that we are following someone else's priorities for our time is connected to the issue of pleasing yourself instead of someone else. Of course, external instructions may at times legitimately trump our own priorities. Helen once said, "Putting out brushfires is part of my job." Anyone with administrative responsibilities has to see the truth of this; one can't always choose, or predict, what problem will take precedence on a certain day. Some of us have been reminded by Group that we are really *good* at stomping fires and realized that to some extent we enjoy exercising that essential administrative skill (though our boots may get singed in the process).

One must be vigilant to prevent planning and prioritizing from becoming a route to overbooking. I began this chapter with a discussion of having everything so planned that we have no time to breathe, to appreciate, or to enjoy life. When I worked on a contract to expand my horizons—that of wanting to read more literature and poetry—we agreed that I had to make the goal to create the space for these activities, not create an expanded list of things I must do. Someone responded that after she made lists of things to do, if there were any gaps she filled them with *more* tasks. "I feel like the rest of my life is polymerized," suggesting that her future seemed to be a fixed array of plans with no remaining flexibility. The point is to let the gaps remain as free time. Leaving some time unprogrammed is itself a kind of control.

A lesson I've learned from our work on time and equilibrium is that slowing down, not working frantically all the time, is not only good for me personally but usually results in my working more efficiently. Time left unscheduled may get used for creative work or undirected thought. Helen observed that "things are different when you move slowly, wait to solve the problem, instead of jumping in to take care of everything fast." And Judith reminded me that "the point is not always to get the job done; it is to enjoy doing it." Paying attention to time and tasks dovetails with our need to make choices and to be in the moment, taking pleasure in the process of what we're doing.

7 *Flying Furniture*

CHOICE AND CHANGE

- *Acknowledge that you are scared, then do it.*
- *If you are stuck, honor it. Maybe you are in the process of making a change. If you make space, something will bubble up.*

"The furniture is flying about the room, and I know that when the wind dies, it will all land somewhere, but I don't know where." I love this metaphor for change and refer to it when I'm describing turmoil in my life. Suzanne elaborated on the risks: "You might have liked it better the way it was to begin with." I countered with "But you must have doubted that, or you wouldn't have opened the window."

We frequently remind one another that change, no matter how positive, is stressful. Helping members through times of change is a basic function of a support group. When lives are in motion, we have provided perspective and grounding. Some of our efforts are of the "egging on" variety, giving someone encouragement for a choice that has already been made but that remains scary. But we focus more on helping others recognize that choices exist and on figuring out how to make them. Changes that are chosen bear an extra element of stress over those that just happen, because the chooser bears more responsibility if things don't turn out perfectly. So it is sometimes tempting to let go of the reins. But if we take a passive role, rather than making choices, we relinquish power and give up the opportunity to create a maximally satisfying outcome. Group members try to help one another make decisions honestly and intelligently, accepting the risks in hopes of getting the rewards.

I discovered how scary positive choice could be when I shifted from personnel work within Cetus to a business position in a new division. Completely my idea, the move was complicated because it involved a demotion of sorts. I would be stepping down from the company's management committee and reporting to someone who had been my peer in the corporate structure. Moreover, I hoped to do this without a salary reduction. The attraction was that the work would be new, exciting, and closer to the science and technology that I had missed in the human resource position. Group encouraged me to go for it, and I got very excited. When I proclaimed, "My gut feeling is positive," Christine teased, "You have been getting complacent about your job. Something has to change." She had known me long enough to recognize my thirst for challenge, and her comment gave me confidence that I was making a wise decision in the context of my own history.

Events moved quickly. I persuaded my prospective boss and top management, and within a month they had agreed essentially to the plan I'd outlined. Having gotten what I wanted, I came to the next Group highly agitated, saying I wanted to "validate my ulcer." I was nervous about the new job and overwhelmed by trying to leave everything perfect for my replacement. Group assured me that my state of mind was reasonable and predictable. I had done a forbidden thing by saying that I wanted the job, would be good at it, and should have it at the right salary. People at work now envied me— and being the object of envy is scary. Perhaps I feared that I had gone too far. Group suggested an affirmation to remind me that positive choice is a difficult thing: "This courageous choice has given me the opportunity to learn about myself and my feelings." Someone reflected that it was sometimes just as difficult to feel terrific as to feel depressed, to which Christine responded wryly, "Certainly less familiar," and my work ended in laughter.

Numerous members of Group have considered new job possibilities, many of which have involved a geographical move, which further complicated their decisions. Most often, we have advised one another to investigate, to assume that we have the right to get information even if we doubt that we will proceed. When Suzanne was pondering a possible move to Boston, where her husband had been offered a prestigious job, we suggested she lighten the psychic load by getting herself invited to give a research seminar at the department where she might want a job. "Consider it an exploration, not a job search. Feel it out from there." Almost ten years later she was considering an invitation to speak that was attached to another university's job search. "Why should I agree to give this interview seminar? I don't think I will have an interest in the job." The answer was, "To see if they can tempt you. They want the chance to do that. That's why they asked you. You don't have to know the answer." This is an observation about limits to personal responsibility. Exploration of new possibilities is empowering, and we can choose to investigate without second-guessing the judgment of those who extend the offer.

Another barrier to making a big change is the fear that all that we've done before will be invalidated. I experienced this when I stopped being a bench scientist. If this was the right thing for me to do, had I done the wrong thing all those years? It seemed that either my future intentions or my past endeavors must be invalid. People who expressed great surprise at my choice exacerbated my dilemma. Not only did former colleagues question the decision, but more than one young person hearing my story of leaving science for business has commented, "Wasn't that a terrible waste of time and training?"

This concern about invalidating the past comes up in other contexts as well. Acknowledging a wish to focus more on personal and family relationships or to consider one's spiritual side may lead to questioning of earlier decisions. Several of us struggled with this in different ways, one describing herself as "scared by the contrast between how I've been and how I want to be. My personal life is empty. I feel like a failure despite evidence to the contrary. I'm fifty and I've given up so much." Christine said, "Growth only comes from recognizing that the old has to change. The choices have changed, the ones you made before aren't wrong." This work laid a foundation from which I could combat the fear of a wasted past when I left my corporate job to write.

Looking at change from the point of view of not rejecting past choices connects to the theme of accepting and honoring ourselves. Self-acceptance works only if we continue to stick with it through changes. In addition to loving ourselves just as we are, we have to trust that if we make changes we'll accept and love the new self we grow into. Changing course becomes unbearably scary if we can't trust that we will be able to look back, even at our missteps, and say with acceptance, "I did the best I could at the time" and even "Now I see I could have done better."

Recognizing where we have choices, then weighing and making them is the most exciting part of change, but we've also found common issues cropping up after a decision is made. One example is the compulsion to finish everything up at an old job, leaving things in good shape before moving on. I do the same thing with home chores before a vacation, feeling the need to write letters and leave the garden weeded and pruned before going away. When I had hired my replacement in the Cetus personnel department, I created impossibly long lists of things that had to be done before I left. Group advised that I call the list "Things the Personnel Director Has to Do" and remember that the title would soon refer to someone other than me. I figured out that part of my urge to leave everything in good order came from the vision of someone else at *my* desk, noticing what I had left unfinished or hadn't done perfectly. (This is an illustration of the Fraud Pig at work; see Chapter 16.) I followed Group's advice and let go of the concerns on schedule. Using my experience to create the list of critical tasks and issues was appropriate; continuing to worry about them was not.

After making a hard choice, we have learned to expect a roller coaster of feelings, some validating, some negative. We encourage one another to experience the feelings, not to dismiss them for fear of having regrets. Mimi made a contract when she decided to stay at Berkeley instead of accepting an attractive job offer elsewhere. "I will not hate myself for this decision," which Group amplified to "In fact, try to love yourself for it."

We enable ourselves to make choices by believing that we are entitled to do so. "Ask for 100 percent of what you want" is a fundamental Group affirmation, but we often include in our contracts the concept *"Recognize* what I want and ask for it." This reflects that knowing what we really want is an even bigger challenge than asking for it. Noticing what is going on is one of the most important tools in determining what we want. "Pay attention to what feels good and what does not" seems obvious, but I know I'm capable of having long periods of just getting things done without particularly noticing how I feel about doing them. Another lesson several of us have shared is that when we create space and take time to notice what gives us pleasure, we often come around to figuring out what we want. Possible results may include noticing what is missing or finding that we already have what is most important to us.

Figuring out what we want and making choices is hard work, and in the end it is about being honest and trusting ourselves. As Helen summed it up, "We have to learn to trust our perceptions; at some level we really do know what is good for us." It is vastly rewarding to examine the elements of one's life, determine possibilities, make a viable choice, and carry it out.

8 *Best Friends, Harshest Critics*

WORKING WITH OTHER WOMEN

- *Women aren't as fragile as you think — our relationships are durable.*
- *It's difficult to be a role model when you didn't have one yourself.*

I still find it odd to hear Group referred to as "a women's group," although we are, incontestably, a group of women. I react this way partly because at the time I joined, the fact that men and women had established Group together was a defining element. I perceived women's groups as involving a certain amount of sitting around and complaining about put-downs from men and the hard lot of professional women. Although this perception may have been unfair, our determination not to follow that path has probably helped us to maintain a positive problem-solving focus and to avoid blaming men in general for our professional problems or the inequities that confront us.

Regardless of Group's genesis, there were men in it for only the first three years, and all but three of the current members have known it solely as a women's group. This evolution did not involve an intentional exclusion of men, but neither was it completely accidental. When we sought to replace departing members with people who were interested in the idea of a group, who wanted to commit to the process, and for whom we sensed Group would be useful, the prospects who came to our attention were women. After considerable discussion, we decided that male representation would not be a priority. I sense now that the intimacy we have achieved and so enjoy is a function of our all being women. Most of the issues we work on are universal, but we gain power and focus from considering how women view and tackle such matters. We sometimes think and talk about how our needs and style differ from those of most men we know.

After Suzanne had been in Group about a year, she said, "I want to figure out why I distrust women. I always suspect that women are being devious. I feel competitive with women and not with men, and my competitiveness is fiercest around traditional women's roles." I noted that she had already made progress in developing trust in women, at least the women of Group, to be able to say this. Advice from Group was that she make a positive commitment to trusting, without trying to figure out why it was so difficult, and that she "pay attention to the bell," noticing the negative feelings as they arose, but not letting them take over.

I don't think Suzanne's revelation surprised any of us particularly, although our individual experiences varied. I had developed primary friendships with women in high school, college, and beyond, and I valued them at least as much as my relationships with men. Nonetheless I was familiar with the kind of intragender competitiveness that seems to lead some women to resent the success of other women while accepting with equanimity the achievements of their male friends.

Our commitment to Group and our interest in encouraging and helping others has prompted numerous soul-searching discussions about why strong, achieving women might avoid or mistrust other women. Many of us went through our professional training in environments where most of our colleagues were men, so we have actually had more practice dealing with men in the workplace than with other women. Additionally, the behavior of some women colleagues has put us off. Many girls, in times I hope are now past, learned from their mothers that a certain amount of deviousness is necessary to achieve social goals ("Don't act too smart"; "Don't act too interested in a boy") and have carried those lessons into their professional lives. Women who have rejected this mode of interaction in favor of straight talk are very sensitized to it in others and react by thinking, "She's not being honest; I can't trust her."

Suzanne's early work on trusting women is the backdrop for one of my favorite examples of rapid progress on a contract. Cooking was one of the traditional roles where she felt a need to be superior, or preferably perfect. Not long after she introduced this work, we had a dinner party. Suzanne made an elegant French chocolate cake, richly and smoothly frosted. On cutting into it we discovered that she'd left the pan-lining waxed paper on the cake, so that in order to eat the frosting we would have to scrape it off the paper. Eyes turned to Suzanne, a little fearful of her reaction, then we all burst out laughing, she hardest of all. When we caught our breath, she said, "I guess I've come a long way if I can laugh with six other women about a *cooking* mistake." It's typical of Suzanne to make quantum leaps toward a goal immediately after deciding to work on it, and the story has become a Group legend.

Another phenomenon we have recognized is that we usually have high expectations of the women in our lives. We want a female colleague to achieve at the highest levels and to be an exemplary person, and if she doesn't meet those expectations, we feel a sense of betrayal that we might not experience if a male colleague fails us. Feeling betrayed is a risky reaction in a professional milieu because feeling betrayed leads to resentment, paranoia, or a desire to retaliate, none of which are useful in the workplace—or in personal life, for that matter.

One Group member had been instrumental in the hiring of a

junior woman in her department who was neither interacting sufficiently with other faculty nor working hard enough to meet the requirements for tenure. The senior woman wanted to help and advise but was resentful that the young colleague wasn't living up to the high expectations she felt for her as a mentor. Coupled with the resentment was the suspicion that she had not been a sufficiently good mentor and was in some way responsible for the younger woman's failure. Group reminded her, "It is not about you. She is responsible for her own success; you can't force her to follow your advice." Someone else reported being very irritated with a scientist who was a sabbatical guest in her lab: "She's well-known but technically helpless. Helplessness is something I really can't stand, *especially* in women." Again, women apply the highest standards to other women. My take-home lesson is this: if I know a woman who is floundering, I have to be as careful not to be too hard on her because of my high expectations as not to be overly forgiving because she is a woman.

As a professional woman, I have felt a tremendous obligation to like and support other women and was troubled if I didn't carry through with this principle. Not liking a man I knew at work didn't make me uncomfortable; I would assume it was because I didn't think much of how he did his job. In contrast, I tended to take my dislike of a female colleague as a personal failing, searching for jealousy or competitiveness as a root cause. I learned from Group that I was not alone in fearing that I had an ulterior motive, and I came to accept that it would be highly un-usual to like and respect all women with whom I interact pro-fessionally. Suzanne had the experience of a senior woman who tried to be her mentor but came across as overmothering and controlling. Suzanne reported that she had been avoiding this person but felt guilty because she had been rude without explain-ing why. We helped her formulate a plan to speak openly and seek a different relationship.

I have mentioned our struggles with the tendency to try to please others at the cost of our own interests. Some attribute this to our mothers' programming, others to behavior we learned in order to survive as little girls. Whatever its origin, this wish to please manifests itself especially in relation to other women. The instinct to judge harshly women who haven't met our expecta-tions seems linked to our high standards when we try to demon-strate our own strengths to other women.

The rewarding side to having female colleagues is the pleasure of interacting with a woman whom one likes or admires, whether by serving as mentor to someone younger, asking the advice of an older, more experienced woman, or sharing experiences with a contemporary. Group has strengthened our appreciation of these relationships and the importance of maintaining them.

Being a role model is a complicated mix of joy and burden. When I had been on the faculty at Berkeley about two months, a female graduate student told me I was her role model. I wanted to protest that I couldn't *be* that on top of everything else asked of me. Yet my credentials were undeniable. I was the first woman faculty member in the department. She had been a student there for four years before I arrived, and she held the wholly justified view that my presence made a difference to her. Group agreed that we needed to have compassion for ourselves in this arena. Many of us had not had any female role models or mentors, so it was not surprising that we didn't instinctively know how to be one. I think Judith may be the only Group member with a female mentor who had a significant influence early in her career.

Group's attitudes toward the concept of role model have evolved. In the mid-1980s, Christine reported, "I am really feeling attached to the women in the first-year class. It's a strong group. For the first time I am feeling okay enough about myself to want to make it better for other women. After hating being a role model, it's now enjoyable to be there." Judith, as an established senior faculty member, told us she wanted to "recruit good women" to work in her lab. "Things are going well, the lab is established and the people are good, but they've all been men of late." The men were capable and were making contributions to the lab, but Judith believed she could provide something special to women as a scientific mentor.

In addition to our interest in ensuring women a place in an organization we lead, Group members discussed the importance of mentoring men as well. Carol, for example, recognizes that her approach to running a lab has influenced male students and postdocs who have gone on to run their own. It would be a mistake to lessen the impact of our unique strengths by passing them on only to other women. In addition, as Mimi commented, a professional setting with men and women working elbow to elbow represents the real world that we hope our students will encounter.

Issues about relating to other women are closely related to our definitions of self. In becoming professionals, many of us have achieved a strength and power that conflict with our traditional images of womanhood. Consequently, we have found ourselves denying the very power for which we've struggled or feel guilty or uncomfortable having such power. Suzanne reported her struggles with self-image from a gender-based angle that struck a chord in me. Having for many years put her career development second to being a wife and mother, she thought long and hard about gender roles as she tried to resolve internal conflict. "I find myself so programmed to dismiss, to loathe the female role, the caretaking, that it is hard to value those traditional female skills in myself, my mother, my friends, or my colleagues. My own self-image is of 'being a boy' in the world. In my experience, 'being a girl' means everyone else has the right to say how you should be." Suzanne's new goal was to redefine her model of femininity rather than reject the feminine.

One person described her difficulty with being positive about her successes and noticed that it was especially acute when talking to Group. She suggested that this came from her attitudes toward women. "I'm more comfortable being my full successful self with men, perhaps because men are more comfortable with their colleagues acting that way and are not threatened by a show of strength in others in the workplace." She had a sense that the way to forge a connection with other women was to be needy, in pain, or unhappy, that women liked those who approached them from a position of weakness so that they could help. This attitude had resulted in her projecting the image of a troubled person compelled to talk only about bad feelings when she was with women. Her immediate contracts were to "watch what comes up and change the habit of focusing on the bad" and to tell us in the next Group about positive events and things that she liked about herself. A long-term contract was "Accept negative thoughts but don't identify with them. Share triumphs." This was fruitful work. Several of us profited from a focus on getting comfortable with projecting a positive self-image.

Group has helped me learn to value and respect the special qualities and skills women bring to the workplace. I started out believing that my ultimate goal should be to have my work environment (the world of science) be blind to gender. After leaving academia, I saw those women who stayed in the field develop a greater recognition of their unique gifts. In 1991, Christine was asked to give a commencement address and was assured that she had been chosen "not just because you are a woman." The students had asked her to avoid gender in her talk, anxious to hear her thoughts as a scientist and professor, not as a "woman scientist." She at first accepted this directive, but then realized that "there are still issues. I'm thinking of talking about the paucity of women in the field. I want to ask what women can bring to the field—humanism, compassion." Titling her speech "A Lab of One's Own," she emphasized the importance of choosing a personal style and the value of diversity. She pointed out that personal creativity carries over into science. Judith agreed: "By trying to fit into the mold you limit our own creativity. I have experienced a feminization of my perspective, learning to trust the voices of other women and myself, and I know there is a strong correlation between our trust of our inner voice and our capacity to be our most creative and productive in science. Group has really shown me the possibilities that come out of this feminine voice."

"The role of the outsider is hard," Christine concluded in telling us of her proposed graduation talk. Our struggles to honor a feminine perspective have been colored by being the minority in our workplaces. As more women enter a field, being an outsider because of gender becomes less likely. But the lessons about maintaining a strong sense of oneself personally and professionally remain important whether or not gender is a defining issue.

9 *Life Is a Limited Resource*

- *Hold less, balance more, and change positions often — mentally, physically, and spiritually.*
- *I'm going to be my own best friend today.*

"I am entitled to take care of myself." Nowhere is the concept of entitlement more appropriate than in this context. It is a constant challenge to pay appropriate attention to our health and our bodies. And when we are going through difficult times, we need to give special consideration as well to our emotional selves. Many of us have neglected self while attending to other people and responsibilities and have had difficulty asking for help that might sustain us. Yet physical and mental well-being should be a top priority in our lives.

It often appears that we say to ourselves, "I'm not yet down enough that I need to take care of myself." Then we up the definition of how sick we would have to be. When I was denied tenure, I fantasized about having a nervous breakdown; then nobody could say I shouldn't get some rest. I was, of course, the real judge of whether Ellen deserved a break, and my answer was always no. I had to soldier on unless or until I became completely helpless. Some years later, I reported feeling totally dragged out with flu-like symptoms, but I had so much to do at work that I couldn't bring myself to stay out of the office for even one day. Mimi said, "If you were run over by a car, you'd stay home." Watching me process that thought, everyone said in anxious chorus, "But don't!" I wanted a clinical explanation for my exhaustion, something medically treatable so that I wouldn't have to slow down or so serious that I would have an excuse to devote time to feeling better. A routine blood test had indicated a thyroid imbalance, and when repeated tests proved normal, I was actually disappointed.

We may not feel entitled to get enough sleep, or stay home with a bad cold, or go to the doctor, yet we still feel responsible for tasks. Someone reported feeling both guilty because she had stayed home with a cold and irked at herself because she hadn't taken charge of her illness, having left work only at the insistence of her colleagues. Group noted that this approach made "even the act of taking care of yourself a chore, another area in which you need to improve." Helen suggested the affirmation "My body is sick; it is time to let my body be sick." Everyone acknowledged difficulty in taking downtime and in accepting a day off for healing as a positive act.

The lesson we need to learn again and again is that by taking care of ourselves we become better able to fulfill our responsibilities. This is especially true of routine exercise breaks or a simple change of pace. Much of my job as personnel director was about being available for everyone else's problems. One of my most useful contracts in that period was "Close the door and breathe. Sit quietly for a short time, or take a short walk." When I reported that I had finally incorporated regular midday gym breaks into my schedule, and felt better for them, Carol said, "Maybe you are realizing that you work more efficiently if you don't work all the time." Years later, writing this book, I have had to learn again that for me exercise is essential to working well, and to omit it is almost always a mistake.

Administrative jobs at the university are also often defined by bombardment with the needs of others. One contract on the subject is poetic as well as practical: "I have thousands of interactions with various people. I want to make as many as possible of them human. In order to do that I have to take care of myself. These things are *more* important than being efficient." A related affirmation is "Recognize that the world won't stop when you stop to take care of yourself." This came after the speaker had been so ill that she'd had to stay home for a week. Of the things that had to be canceled, she reported, "few made a difference, and even fewer were things I wanted to do." She resolved to take better care of herself in fulfilling obligations, for example, by making lunch dates with special people to balance her more onerous appointments.

Administrative work highlights another aspect of taking care of ourselves on the job—for example, how to be less emotionally invested if a candidate accepts a job offer or if someone is disappointed. "How can I be sensitive to people's needs, which makes me do my job better, without being sensitized to every result, which kills me?" Those with experience in managing people in distress shared what they'd learned: "Desensitization is a process. You learn that the cost of feeling everything is too great. Don't worry about losing sensitivity; worry about taking care of yourself. In administration, caring what happens while protecting yourself frees you to make decisions optimally. Keep in good shape. Managing your own resources is the key to making decisions."

This effort has not been as straightforward and obvious as it may appear. We flounder in our attempts to take care of ourselves. More than once someone has reported feeling guilty for having taken a vacation "just when there was lots of important work to do in the lab or office." But when is there not a lot of important work to do! The vacation is necessary. Another pitfall is to forget what *self* really means. After listening to Suzanne's report of stress, overwork, and efforts to balance home and office, Judith pointed out a missing element: "Time for your family is not the same thing as time for yourself, and you need both. Time for Suzanne is most important right now." We have all echoed this need to distinguish "time not at work" from "time for me." When David and I worked at the same site, I began going to work every day on his (very early) schedule. Then I found I could rejuvenate myself by staying home until 8 A.M. some mornings. Even an extra half hour alone before starting a ten-hour day at the office could make a difference in my energy level and sense of well-being.

The best way to take care is sometimes to ask for help. Group members frequently offer to help one another during crises or times of stress. The big challenge is that we aren't comfortable accepting assistance. Three reasons why it is hard to ask for help are the desire to demonstrate strength and competence by "doing it all," the fear of being exposed as excessively needy, and the fear of being a burden on others. The first emerges as a need to be able to handle everything and as guilt if someone else does something for you. Group developed some different ways to think about this issue: "Identify the times you feel the compulsion to take care of something. Who is talking inside? Are you compelled to fix everything yourself because you value only the things that you have to work hard for?" Another approach to encouraging people to ask for help was to restate some fundamental precepts for working with a group: "Not letting people do things for you is keeping them out. If you do it all, then the other person can't do anything." I proposed the possibility of fearing that if I asked for something, someone would disappoint me by not doing it. The response to that is, "But then you take responsibility for everything and you are disappointed in yourself if you can't do it."

An illustration of the second barrier to asking for help is Helen's personal script: "If people take care of me, they'll find out what a phony I am." For Christine (in the mid-1980s), asking for help meant being needy, the "identified patient." Having recovered from a place of extreme need, she continued to equate asking for help with being sick, "not fixed." She denied herself permission to call or ask Group members for help between meetings. As a founder of Group, with the most experience of the process, she was so identified as a source of wisdom and insight that asking felt wrong. But, we responded, "if you don't call or ask for help because you know more than others, it implies that what you need is an answer. Instead perhaps you need to just be in touch, to share feelings." Christine realized that she was out of practice. "I can't remember the last time I called someone because I was freaked." Affirmations came fast and furious: "Asking for something for yourself is not being in a bad place; it's being in a *good* place." "You need to deprogram the belief that you are weak if you call." "It's about acknowledging your feelings, not about being helpless."

Turning to the third obstacle to asking for help, the fear of being a burden, we must rely on our friends to take care of themselves when they respond (e.g., avoid rescue). Only if we trust that others will limit their assistance according to their own needs can we freely ask. A typical contract could be "Ask for what you want from members of Group. Don't refrain from asking because you know someone is busy or sad; it might be that helping you is just what they need. Or, it might not and then it's their responsibility to tell you." A big advantage of asking Group for help is the number of candidates, so that the request is spread out. One mini-contract read, "Call the first name on the list. Subcontract: Call the second name on the list."

One complication of asking is that calling for help or reassurance often means making an actual phone call (all of us have slight aversions to telephones). Some felt they were frequently an interruption; others resented the distance between parties, wishing that there was time to meet face to face. Christine believed she had to be in good spirits to phone, because her mother had always demanded that her calls be cheerful. The advent of e-mail has helped those who are reluctant to call, but we've also made efforts to increase phone contact. Suzanne, one of the most phone resistant, made a contract to "enrich my friendships and acknowledge the phone as a means to that end." She got a cell phone and started calling when stuck in traffic on her daily commute. Since I retired, Helen and I have touched bases in the early morning several days a week. While I was writing this book, Carol sensed that I needed more outside contact in my lonely home office and began to check in occasionally. Now, either of us may call, and each of us has permission to say, "Can't talk now," but we value the regular contact. When she catches me in the middle of writing, I often talk out whatever puzzle I'm trying to solve and return to the computer with a clearer mind.

From time to time, we step in to help one another with big tasks, like moving to a new house, or routine ones that have become overwhelming. When Carol was involved in writing her book, her domestic mail piled up unopened. Mimi and Christine developed a plan to join her, one on Saturday and one on Sunday, to open the accumulated envelopes. At the next meeting, Carol reported that life seemed easier, back on track. Those few hours of help had broken the logjam. Accepting this kind of assistance may make a personal job that seems onerous quite bearable and even fun when done with or for someone else—a useful perspective indeed.

With some practice, we can even truly enjoy being helped. After Helen broke her arm during a Group trip to Washington, D.C., she reported, "It felt good to let people take care of me." She called it a triumph over her pattern of disconnecting from everyone when she was sick, especially since she'd retired. The rest of us suggested that she make a contract to ask for something the next time she was sick—food, a visit, or a phone connection. We all agreed we could use practice asking for help and admonished Helen to "remember you are doing us a favor because we're caught up in our own affairs and enjoy being pulled away for human contact."

There is a connection between learning to ask for help, letting others take care of you, and self-acceptance. If you can like yourself and not demand perfection and strength, then you can also ask for and accept admiration and friendship much more easily. "It's about letting people in instead of winning them over," Helen said to me when I talked of having to prove myself all the time by doing everything perfectly. "Believe you are okay, and that others will recognize that."

Taking care of oneself is a necessary commitment, a matter of common sense, and an amazingly hard goal to maintain. A special joy of Group has always been the opportunity to take care of the others. My pleasure in doing so should remind me that others enjoy caring for me as well. Helen once said, "Think of it not as asking for help, but as letting us share your life." We keep an eye out for one another's needs, while remembering that each of us has the primary responsibility for herself.

10 *Permission to Feel*

- *My feelings are my feelings; they are not who I am.*
- *The thing is to have the heart to let it all in.*
- *Give myself space to feel. Feeling is a legitimate use of my time.*

We use Group to describe and experience difficult feelings. Sometimes one of us starts with the objective of understanding how she feels about a situation; other times unexpected emotions surface in discussion of a practical issue. The key lessons are that we have a right to our feelings and that disregarding or denying them is counterproductive in the long run. Honoring our feelings is complicated by the existence of circumstances in which it is better not to share them—a tension we continually recognize and evaluate.

In the early days of Group, I was obsessed by a sense of inferiority and ambivalence that led me to question whether my desire to do science was sufficient to the task of being a professor. I introduced my work with the goal of suppressing those emotions so I could plug away at the manuscripts I had to complete. Group response was clear: "To achieve change you have to live through the feelings. A certain amount of time is necessary to achieve that, so all the minutes you spend on your terror are valuable."

When I was denied tenure, the first piece of advice for coping emotionally was "Experience your feelings as they come, when they come." In subsequent months, as I felt desperate and scared but tried to be stoical, Group added, "Let yourself be sad alone. Give yourself permission to feel what you are feeling." A suggestion that seemed to fit was "You are afraid to let yourself go because of the risk that you won't come back. You might get crazy and always be miserable. That's not true, you will come back, and you *need* to go there." Again and again during that time, I went to Group meetings ashamed of not yet having recovered and came away believing that it was okay (even reasonable) to be depressed and unhappy for a time.

The concept of permission to feel is expressed in others' contracts too. Why do we need permission? Why are we suspicious of our feelings? One possible reason is that early on we were all conditioned to view strong emotion as contradictory to professional success. It is as if we think there isn't time for strong feelings, good or bad, that they will distract us from the tasks at hand. As we have moved away from this mindset, we recognize that we need to be selective in choosing those to whom we express certain emotions. We have reminded one another that even though it is sometimes better to hide our strong feelings from others, we must acknowledge and own them internally. Group can be a safe place to explore feelings that we want to keep private.

Several of us shared a fear that sadness is a sign of weakness. Christine captured the essence of this fear when she said, "There is a part of me that is fundamentally sad, and that's different from being depressed. I'm used to equating sadness with depression, so when I'm sad, I become resentful and angry with myself. But there is a distinction." We developed the thought that with sadness, we can get through it by staying with it. I added, "Don't try to explain why you're sad, just be it." I reflected on the joys of being sad at a movie: "You have an excuse to cry, which feels good." Of course, the tears may be about something entirely different from the movie. Through Christine's work I became aware that I never allowed myself to be sad, and I was inspired to explore that side of me in a subsequent meeting. "I can't accept the side of me that might be sad, unhappy, worried, sick, not strong." To which Christine responded, "You plus me would be one hell of a well-adjusted person." After the laughter subsided, Group helped me attack the message in my statement. Christine proposed that I saw vulnerability as synonymous with weakness and needed to distinguish the two. "Having feelings doesn't take away power. In a sense, strength is letting the feelings in." I crafted an affirmation: "I have a choice about when to be vulnerable and when to protect myself."

Group talked about the various ways we have of denying feelings. Most of us were skilled at hiding from feelings by being very busy, by living in a "detail cocoon" that leaves no room for emotions. Many of us grew up with the belief, perhaps instilled by our parents, that emotions need to be kept in check. We are trained to pay attention to our minds over our hearts, and the mind tries to avoid sadness. I reflected that my mind always argues with any sense of sorrow, saying, "Look at what I have. How can I be sad?"

Suppressed or unacknowledged bad feelings about something very specific often spill over into other parts of our lives. If I'm frustrated because I've taken hours to straighten out a wrongly denied health insurance claim or a bank error, the sense of powerlessness colors everything else in my life. I once cried for an hour on a plane, ostensibly over a disagreement with our company's travel director and a flight cancellation that made me late to a meeting. Then I realized that my tears were really for an old friend who had died recently. I had not seen him in several years, and I had not taken the time to mourn his passing. My need to cry had simply caught up with me. Experiencing bad feelings and identifying their sources is the best way to keep from generalizing them. Furthermore, it is hard to selectively suppress only the bad feelings, so we miss out on the chance to celebrate and acknowledge the good. One of us made a dual contract to "experience pain and focus on the good stuff," because she felt she was shutting both out together.

Another reason for probing our feelings, maybe the most important one, is that examining how we feel gives us clues as to what we want to be doing, what is right for us. When I was stewing about my retirement decisions, Group reminded me that to squelch the feelings was to deprive myself of information. "Be curious instead of afraid." I made a contract to "recognize the value of my feelings (excitement, pride, anxiety) at this time of transition and give myself permission to experience them."

Bad feelings can be especially disturbing when they are familiar or revisited. One woman reported being upset about a faculty meeting in which she had felt disrespected. But even worse, she thought she had gotten over the reaction, so this resurfacing felt like a regression. She was reminded, "Accept your feelings. Have the feeling, but don't act on it. Get a sense of space." This brought back a point we'd discussed before: "Reflecting an old pattern, one that is hard to break, doesn't mean you are the person who had those feelings before."

Group has discussed grief and has shared our experiences and a host of books on the subject. As with other emotions, an important theme has been choosing to live with grief rather than suppress it. On the death of my favorite aunt (my mother's sister, with whom I'd been very close as a child), I told Group, "I don't know whether or how to mourn, as though there is a disconnect between my brain and heart." Mimi, whose father had died earlier that year, pointed out, "When someone you love is not part of your day-to-day life, there are few reminders that they are gone. There's nothing to make it real when the person lived three thousand miles away." As I planned a trip to Connecticut to help my mother clean out my aunt's house, I still felt no emotion and resented my job for keeping me too busy to sit back and let the feelings come. I made a note that "none of Group is worried that I won't be able to cry." Mimi advised, "Don't have great expectations about feeling, or crying, or having your mother open up to you about *her* feelings—just go, and be." It worked; the house got cleaned, and my mother and I cried and laughed and talked about my aunt.

Helen discussed how she felt when she was forced to move her dying mother to a care facility. Aching over her mother's incapacity and confusion, Helen experienced a deep sense of loss that her mother was no longer in their family home. Helen saw everything in her life changing and envisioned herself as being "in molt." She spoke of using drawing and sculpture as an outlet, "something to express emotions without being intellectual." Someone commented that "art is a conduit for the feelings, to get access to special energies at this crisis time." Building on the conduit idea, someone added, "like a volcano, getting access to something inside." Combining volcanoes and molting, "Molten Molt-ing" rolled off someone's tongue. Mimi pointed out that when an insect molts, it makes a new, much bigger case but doesn't have to grow to fill the case yet. It just has more room to grow at its own pace. This is a lovely way to envision change and making room for a new set of feelings.

Anger is an emotion that poses some special issues. We want to acknowledge that we feel anger, but it is highly charged and can be debilitating because it may take over and prevent us from moving forward. Helen quoted the Buddhist Thich Nhat Hanh: "When we are angry we are not usually inclined to return to our-selves." The tension between honoring our feelings and suppress-ing them can be especially acute with this emotion. We also know that expressing anger toward the one at whom it is directed is often counterproductive. To this dilemma we've directed contracts such as "I have permission not to act on my anger." In a general dis-cussion of anger, someone summarized, "Fighting with another person to get through stuff can be good, because you are stand-ing up for yourself. Fighting to undermine (or put off) the other person is seldom good." We admitted that this formulation might not be useful, because the object of the anger may not recognize the intended purpose.

We have worked on the sense that anger is inappropriate, as though it's not the emotion we *should* be having. This perception may come in part from childhood lessons that anger is not a nice emotion, that our friends, perhaps even our parents, would not like us if we were angry. This is an especially likely lesson for girls. Mimi quoted her mother: "If you get mad at Virginia, she's never going to want to play with you again." And more cosmically: "If you get angry no one will love you." I often talk myself out of anger. Just as I deny myself permission to be sad because of all the wonderful things in my life, I am not allowed to be angry about something because I start listing all the pluses in my life. This is an issue of entitlement. In a way I'm saying that as long as I have good things going on, I can't be angry about a wrong done me. We need, instead, to own both the good and the bad.

Connected to the sense that we mustn't be angry is the feeling that it is unacceptable to have anyone mad at us. Mimi worked hard on this issue in the late 1980s. "Much of what I'm trying to do is aimed at preventing people from being angry with me. I'm motivated less by making people happy than by avoiding their displeasure. My fear of anger is worse than the pain of not getting what I'm entitled to." She was in a struggle with people in her lab, who were mad at her for claiming her own lab equipment for experiments she needed to do. As Christine pointed out, "We see anger as an endpoint, not as a natural emotion." Mimi's resulting contract was to try an experiment: "Let people be mad and notice that I survive. See if having them angry will feel better than not being entitled to take care of my needs. Honor the choice of action that leads to their anger." Mimi elaborated on this work in the next session. "I never give others a chance to be mad at me because I'm filling all the space with apologies."

As with other feelings, anger can be a useful warning signal that something is going on that requires our attention. Suzanne reported that she had thought she was to blame for her frequent fights with a collaborator, "but in fact he puts me down in front of my students, and they notice. I deserve to be mad." Group was quick to agree, advising her to "take sides with your feelings. Observe the warning signs and stop his put-downs before you get mad."

We also work on deflecting anger and identify situations in which anger might be detrimental. One example is becoming upset about the difficulty of getting things done at the university, which we characterized as a waste of energy. "Anger should be against someone who is trying to screw you. The anarchy of the university dumps on everyone; it is not trying to screw you." Anger is sometimes a useful private emotion, a source of energy to deal with a problem. Yet at the same time expressing one's resentment to colleagues may be perceived as asking for special attention. ("I have been mistreated!")

Despite our focus on letting all feelings in (or out), we acknowledge that, at times, deferring them can be healthy. I call this work "permission *not* to feel." Although in the months following my denial for tenure I worked through many issues (including much anger), for years I delayed admitting how hurt and betrayed I felt. Someone suggested that I had "a highly developed skill of protection by denial" of hurt, and I suspect that it was not a bad thing in this instance. I needed to be assured of my competence before I could look at that rejection and experience all my reactions. After Helen's mother died, she performed the chores related to the estate bit by bit, as they felt right, and as she could handle the accompanying feelings. After Carol's husband, Hatch, died, she adjusted to a new life and new way of being, asserting, "I'm in a flat, emotionless state, but it's okay. It's not forever." Carol knew she would get to the feelings when she could handle them.

There is an anecdote that seems an appropriate ending for this subject, an external validation of our acknowledgment of feelings. After I had started writing this book, Helen and I attended a party where we encountered Mina Bissell, a former colleague and Distinguished Scientist (this is her actual title) at the Lawrence Berkeley National Laboratory. Mina had last seen us when we gave our Group presentation at the cell biology meetings, where she had spoken during the question-and-answer period about the importance of the support we provided one another. "It is the most moving testimony I have ever heard. I cried. And I want you to know I could have used a group like this myself and I'm envious. You should get your write-ups and let people have them . . . and also write a book." She was still choked up as she spoke. The story Mina shared four years later was that a talented young scientist who had been weighing several opportunities for postdoctoral research chose to join her lab after hearing these tearful comments. She had made her decision because she felt it would be wonderful to have a mentor who was a good scientist *and* had the courage to cry in public when moved to do so.

11 *Boss, Mother, Friend, Role Model*

WORKING WITH STUDENTS AND EMPLOYEES

- *Remember to care.*
- *Keep your sense of humor.*

Many Group sessions have been devoted to our relationships with the students, postdocs, and technicians who work with us. Maintaining a committed, interested, and well-functioning staff in a laboratory is as challenging as developing a research program and building a scientific reputation. When I left academia, I found that many issues arising with students apply in other contexts as well.

We frequently ask, "What do our students want from us, and if we knew that, would it be what we can and ought to give them?"

Several of us undertook our faculty responsibilities with a goal of having a supportive lab environment, one that, in Christine's words, was "nurturing for me as well as for the students." But it is not always appropriate to rely on students or employees for emotional support. We struggle with our own and the students' unrealistic expectations that we be the ever competent boss or surrogate parent. Christine described the trap of setting such high hopes: "I am the boss, so I need to be okay, on top, always optimistic, available. I hate that part of the job, so whenever I don't feel okay, I avoid the lab. Then the people in the lab resent my absence and I feel guilty."

One of my earliest notes from Group was advice about dealing with the people in the lab when I felt stressed. "When you don't have time to do more, at least let them know you care about them, but are too overwhelmed to give them what they need. Say whatever is *true* about your reasons." This has continued to be relevant to various people's work. Christine told us of trying to be responsible to her students while coping with the attempted suicide of a scientist friend. "I couldn't deal with the people in the lab. One student clearly needed to talk with me and I felt guilty avoiding him. Then I admitted that I was upset and he was fine." He could wait because he knew his needs would not be ignored indefinitely. Following this principle several years later, Christine chose to tell two women in the lab about a crisis in her personal life. Their response was supportive. "One already knew; the other was relieved to know I have a relationship. They need me to be there, not perfect, just there!"

Students sometimes seem to want reassurance or direction beyond what we feel capable of giving. One of us reported that her students seemed terribly needy after she'd been away for a year. "I'm back from a wonderful sabbatical and my students are in crisis, asking questions like 'Should I really be a scientist?' I'm feeling petulant and want to say, 'Why don't you leave me alone?'" Group suggested that she "curb the negative attitude, but baby-sit less. Be up front about feeling sad that your sabbatical is over. Warn them, and use what you learned on sabbatical to create boundaries."

As we deal with students' dependence on us, we've also addressed ways we depend on them. A professor may feel driven to take an additional student even if there are reasons not to do so. This impulse derives partly from a sense of status in being chosen as an adviser, especially when one is starting out. When I began my faculty position at the midpoint of an academic year, I accepted a student who had asked to work with me and was waiting for my arrival. I would have been better off getting started with just my new technician, waiting to choose among students with whom I had interacted. It was a costly misjudgment, because the student was morose and discouraged others from coming to the lab. I didn't have Group to advise me then. Others have dealt with variations on this theme at more senior stages of their careers, as when someone planning a sabbatical did not want to take new students the next year because she would be unable to guide them; yet she also reported being frantic that none would apply. Talking through the situation with Group helped her shake off the paranoia and get clear what was best for both her and the prospective students—and then stick with it. The burden of attracting students is exacerbated at the time of faculty presentations to new students, a tradition in many academic departments. Several Group members have worked on the intense pressure to perform and the resentment at having to market oneself and one's research program. The advice emerging from these discussions was to let the science speak for itself, to focus on the research, and to forget the recruitment aspect. "Give the talk, communicate something beautiful, but don't make it a sales job."

Giving criticism is possibly the most difficult part of the adviser-student or boss-employee relationship. Group has focused on the usefulness of criticism, rather than on how painful it can be for the recipient, and on how to criticize more effectively. When Christine was trying to figure out how to give constructive criticism on a scientific matter, Mimi suggested, "Don't force yourself to have all the answers. Explain the problem to them and encourage them to propose a solution." Group has turned this same advice to many situations. As mentors we are concerned with how to criticize without dampening enthusiasm or discouraging talent. Getting the subject to participate in the strategy for improvement is often the key to accomplishing that goal.

Difficulty giving criticism and being a good boss often results from one's experiences or insecurities. We've reminded one another that we do a disservice to both the students and ourselves if we fail to be honest about their abilities. In addition, we've questioned why we delay in giving honest information. In one example, someone worked on the painful process of kicking a graduate student not only out of the lab but out of grad school. "I should have told this student to leave long ago. He has had problems from a failed prelim exam to the present, but I've delayed this process because of my own paranoias, my fears that I hadn't prepared him well and that the project wasn't good enough. I tried to fix it but I never got honest." She realized, talking to Group, that she had been acting on an assumption that everything was about her, that she had all the responsibility and the student none. The student did leave and expressed relief in finally being able to give up the battle.

Group functions as a place where we can acknowledge irritation and personality conflicts, elements that we have to keep out of a discussion with a student or employee. One member described a student's serious mistake that destroyed months of work by others and why she found it hard to talk to the student about it. "She's a manipulative type, and I don't like the way she uses her femininity, but I can't let that be my focus. I have to figure out how to get her to shape up while feeling so negative about her personality. It's time to tell her that she needs to run her own life, her own science." Having worked out her negative feelings in Group, the professor was able to devise and carry out her plan of action, giving the student a deadline for publishable results in a purely professional context.

Sometimes just describing an issue to Group renders a crisis less fraught with tension. When Christine looked for advice on dealing with "an upset and territorial postdoc causing havoc in the lab," she got serious feedback and a humorous perspective. "Put boundaries on what you are responsible for. You can't make everyone behave." And "be glad they're fighting. It beats having them all at the beach." Christine said her Bad PI (Principal Investigator) Pig was on the attack.

Although we strive to avoid the mother-caretaker role, there are times when a mentor has a real responsibility to step in as protector, a role that may be awkward. When one of us learned that a longtime (male) friend of hers had made unwanted sexual advances to a female student, Group was unanimous in advising her to tell him that she knew about the incident and was angry. We also agreed that a member needed to confront a colleague who had agreed to write a letter of recommendation for a postdoctoral fellow, and did so without telling the applicant or the adviser that it would be negative. These were difficult situations in which speaking up on behalf of a subordinate was a matter of responsibility and principle.

Students do not hesitate to judge their professors. In sharing these sometimes painful judgments with Group, we have found comfort, advice, and humor. One faculty member's students walked out of her lab group meeting because she was aggressively critical. She pondered whether it was possible to keep control, say what needed to be said as a leader and teacher, and still be nice to people. Christine's students told her, after she gave a commencement address in which she talked about the importance of mentoring, that she wasn't a mentor any more. "They say I'm not there for them." These incidents feel like torture, but the frank criticism keeps us on our toes and sometimes inspires changes. Other times we have to just commiserate, compare notes, and know that we can't fix it all to suit the people who work with us.

Relationships with students and employees connect to issues of honoring a personal style and acknowledging the right to have things work the way we want them to in an organization we direct. We have to learn to exert power in an effective way; wishing to be liked may interfere with that process. We have frequently reaffirmed that to use power appropriately we must accept our responsibility, as well as our right, to lead. Mimi complained that people in her lab weren't taking care of shared equipment, which angered her. As a solution, she wanted to change the dynamic. The feedback from Group was "Forget how you feel. Remind yourself that this is your lab. You are graciously letting them participate, but you are in charge. Say, 'This is the way the lab runs. It's my lab.'"

On another occasion, Mimi noticed how much she enjoyed her students when she got to work with them at the summer research station at Friday Harbor, Washington. "We interacted about science, and it was great. I want to re-create that in the lab at home." She made a contract: "To set aside time and space in the lab to work and talk with students." Her one day a week in the lab attending only to research became sacrosanct, and is often the best day of her work week.

We don't stop being concerned about our students when they are no longer working with us. We have discussed worries about former students getting jobs, getting grants, and getting tenure. Sometimes we see ourselves in our students and relive our own struggles as we watch them progress. We've made many comparisons with raising children, whom we similarly have to advise as we can—though we must also let go. Suzanne shared her concerns about an excellent former student, now an assistant professor up for tenure. She wondered what more she might have done to help. Among the responses from Group was "We need to let go when they fail as well as when they succeed. Let her know you love and value her, and let her proceed."

In the end, there are tremendous rewards from working with students and others whom we mentor, and from time to time we bask in those moments. Here are a few favorite examples:

- *A brilliant young scientist in my institution told me, "I want to work with you; that's why I'm here."*
- *A former student got an award, and she called me before telling anyone else.*
- *I spent all day in the lab with a good student. It was a delight.*
- *I received a wonderful note of thanks from a student leaving the lab. It never occurred to me to expect one, but it means an enormous amount.*

Interacting with students as classroom teachers is different from mentoring them in the lab and has its own challenges and pleasures. Teaching is time-consuming, often terrifying, and sometimes rewarding. The completion of a well-designed course, or even a single good lecture, is a great satisfaction but can deplete the lecturer's energy, as Carol reported when teaching two courses at the same time, one to dental students and one to PhD students. "All my integrative faculties are shot."

Perhaps the hardest thing about teaching is to *not* try to please all the students, an impossible goal in any event. We have to accept that we can't make students want the information. If we try and fail, it makes us hostile toward them. Introducing a bit of humor into the discussion, someone suggested that when students don't laugh at a joke or get a beautiful point, the lecturer should think, "Pearls before swine," and move on. Much of the advice we've shared is summed up as "Don't seek their approval. Just teach as you feel it should be done." The added advice to "Care about them" seems simple but is easy to miss. A set of rules for getting through a series of lectures to medical students who didn't seem interested in the subject matter elaborated on these concepts: (1) instead of getting them to like me, try to like them; (2) keep my efforts limited, as appropriate to the task; (3) keep a sense of humor; and (4) hang on to perspective.

In the classroom, as in the research group, students are harsh critics. In 1985, at the beginning of a lecture, Christine found a list taped to her microphone; it was entitled "Seven things you did wrong in the last lecture." We applauded her for getting through the ensuing hour, saving tears for later. Her male faculty colleagues were more amused than distressed, one saying, "Well, if I'd gotten a note like that I wouldn't have *cried*." The graduate teaching assistants in the course, men and women alike, were wonderfully supportive and sympathetic, which we hoped boded well for future generations. Three years later the class applauded her series of lectures in the same course.

Because teaching a major course can be totally absorbing, we've often had to develop survival strategies. Some of us have tried to set up the teaching period to exclude all other obligations, but usually teaching is a series of choices and compromises. We find it hard to accept that we can't know everything about the subjects we include in our lectures. As Christine became comfortable with teaching, she acknowledged that previously her efforts had been dominated by insecurity. "If I can be authentic about what I am, I need not worry so much about what I can't do, or don't know. Fear of the limitation is greater than the limitation. Now I'm finding learning is easier, I'm understanding things that were difficult when I was scared." In a similar vein, someone else described being "terrified of teaching, but hopeful. I stayed with the terror for two days and it wore off, and now I'm enjoying all this interesting material."

These comments exemplify one of the best things about teaching: the opportunity to learn. In preparing a lecture, one often learns new facts or envisions a new way of putting things together, or both. In trying to make a subject understandable and compelling for students, a teacher may come to comprehend it better and find it more compelling herself. To share information with students, in either the classroom or a direct mentoring situation, is to have a partner in learning, because students' responses and questions are an independent and often enriching input to the process.

12 *Putting It Out There*

- *Love the ideas.*
- *Have the courage not to know.*

For most of us, formal communication is an essential part of the job, and we are rewarded for doing it well. Communicating effectively is difficult, however, so Group work frequently focuses on writing papers and giving research presentations. Sometimes the issue is confidence, in that speaking and writing call up the question of whether we believe in ourselves. But the challenge is also organizational and, almost always, motivational.

In 1981, I reported trying to write a paper and "thinking so much about what others think that I can't remember what I have to say." This response is not unique to scientific papers; as I create this book, almost twenty years later, I get paralyzed by wondering how the rest of Group and other readers will react to each paragraph. This work brings to mind the challenge of paying more attention to what pleases me than to the judgments of others.

Why is it so hard to get started on a writing project? Ultimately, the end product is a public presentation of our accomplishments and ideas. We know we are giving others the opportunity to criticize us—and that they will do so. Envisioning that criticism makes it hard to get started. Most of us first write a rough draft, on the principle that it is easier to rework a bad draft than to put the first words down. It is hard to begin, because we don't view our writing as a draft; we think of it as the piece of perfection we eventually want to produce. In one brainstorming session on getting over writer's block, someone advised, "Tell yourself you are just 'catching thoughts' instead of writing." In a lighter vein, though not beyond possibility, someone suggested writing "on a paper towel to convince yourself it's not permanent, nothing you might be judged on." After completing the first draft, we still have to determine when to stop revising and submit the paper, which sometimes means stopping short of perfection, perhaps at the second draft rather than the tenth. Christine proposed an affirmation for deciding that a paper is finished: "This is not the best piece of work I am capable of doing. It is the piece of work I'm going to do."

The awareness that writing requires long uninterrupted stretches makes us want to complete other obligations before we start. There's nothing like a big writing project looming to make us notice other things requiring our attention—from personal correspondence to remodeling the office or kitchen. But we must schedule time to get started on that research paper and let the other tasks wait, knowing that we won't stick to our schedule perfectly. Judith commented, with a positive spin, "When you have richness in your life, you get interrupted."

Another inhibitor is the tendency to treat writing as an add-on to regular work. Although papers are an integral part of research science and crucial to career progression, one often tries to write them nights and weekends, because so many other demands fill the workdays. When I complained that I had been procrastinating on a writing project and dreaded giving up another weekend to it, Group advised, "It is work. Do it at work. Set up a day when you will finish it—and do. And if it's not finished, set up another day." Writing can be a great joy when one finds the time and energy for it. Several of us have gotten long-delayed papers written during sabbaticals or even on vacation, when a little quiet time goes a long way and doesn't necessarily interfere with the holiday.

The obligation to "write it up" sometimes interferes with enjoyment of the research. Writing a paper or talk is so costly of one's time and so associated with fear of criticism that we must remind ourselves of the rewards. These include the opportunity to focus on ideas, to think, and perhaps to uncover a new scientific puzzle. Several of us have had this experience with review articles. These long papers are an overview of an entire field or topic that meticulously references the work of others. The writer must assess the literature, select which papers must be referenced, and synthesize various discoveries to tell a coherent story. It is often a compliment to be asked to write a review, and there may be an element of obligation to the scientific community, but such papers entail a tremendous amount of work. Mimi announced, with surprise, that she was enjoying writing a major review. "I'm learning new things, permitting myself to enjoy the doing, not just the finish." Only the pressure of a deadline diluted her pleasure.

Although we work on whipping up enthusiasm for obligatory writing projects, we also discuss the right to pick and choose what we will write. In the case of an invited piece, such as a review, Group members encourage one another to assess the time and effort necessary to complete the project before agreeing to do it. The writer should perceive an intellectual reward in the task, such as an opportunity to read the literature of the field carefully or to explore a new area. Anyone is entitled to decline an invitation to write a review article if she doesn't have time, is not interested in that field, or doesn't believe that the journal in which it will be published will reach an important audience.

Most pieces of writing are criticized, both before and after publication, and sometimes unfairly. Reporting these experiences to Group helps us pick up and move on. I once received a letter from a journal declining to publish a paper I had submitted because it was "not of sufficient general interest for our journal." An issue of that very journal arrived in the same mail with a competitor's report of a similar study, no more general, with conclusions based on results less clear-cut and less innovative than my own. There was nothing I could do but submit mine to a more specialized journal, grumbling as I did it.

Christine was blindsided by a review in a major scientific journal that panned her chapter in a book that was otherwise favorably discussed. She had agreed to contribute to the book at the urging of other contributing authors, who felt that her unique approach deserved a separate chapter. The reviewer, however, called the tone of her chapter arrogant, in that it focused only on the work of her own lab. Christine was enraged and hurt, asking, "What do I do that elicits such a response?" I pointed out that "if you are good and have a good reputation, some people will call you arrogant. It can't be helped." We also indulged in some commentary about the "self-important small-minded academic" who wrote the review (though none of us knew him). Mimi gave Christine a stroke "for having assumed in advance that the review would be complimentary."

One of the responsibilities of an independent scientific investigator is to apply for grants from agencies such as the National Institutes of Health, National Science Foundation, or American Cancer Society to support the work. The process involves proposing the body of research one intends to do over the grant period (usually three to five years), asking for the money to support it, and preparing a budget to back up the request. Grant proposals are therefore a special category of writing, because so much is riding on the result—a technician's salary, lab supplies, or a piece of equipment that will make a whole new set of experiments feasible. That a grant is about asking for money pushes buttons about entitlement: "Why do *I* deserve these funds?"

In figuring out how to get through the stress of grant writing, we've emphasized using the process to think through projects, to "be creative, not just justify funding." Christine advised, "Make a positive visualization of what you want, how it will look. Create a series of positive statements; look at them again and again." Mimi reported when writing one grant that she was excited about the work she was proposing but was so anxious about the low levels of funding in her field that she couldn't write. The first suggestion was to "get so tired you have no energy for panic," then start. Everyone added advice: "When you are completely panicked, skip to a mechanical task, like listing the references, to calm you down." "Don't be afraid to shut yourself away and just do it." We've discussed the rewards that may accrue if one can just get started. When someone introduced her work by telling us she was writing a grant summary with a five-year research plan, I expected she would work on how to get it done, but instead she said, "It's exhilarating. I love it," and highlighted the big plus of grant writing—the chance to think about what you have done and want to do next.

Carol has completed the grandest publication achievement among us. For several years in the late nineties she labored over the completion of a book that Harrison (Hatch) Echols, her husband, had conceived and begun before his death. This treatise is an intellectual history of molecular biology, written to show nonscientists how scientific investigation is actually carried out and to share the complexities of discovery and the excitement of science. Fulfilling her promise to Hatch to finish and publish the book was a major objective in Carol's life at the time she joined Group. She also was establishing a new lab at UCSF after her move from Wisconsin, learning the politics of a new institution, and taking on new teaching responsibilities. The enormous time commitment required by the book competed with other things she needed to do. It was alternately frustrating and exhilarating, and often a focus of her Group work. Writing a book was not something Carol had wanted to do. She reported, "I'm discovering how hard it is to do the art" (in this case, illustrations elucidating complicated molecular processes). Then a month later: "I'm really liking the work on the illustrations, and I'm getting something out of doing them."

Some of the support Group provided Carol was about prioritizing; sometimes teaching responsibilities took precedence over the book for a time, and sometimes the book demanded that she neglect other obligations. When she dealt with the profound emotional issues, Group's function was mostly to listen and sympathize. Within two years of Hatch's death, Carol was recovering—not forgetting but establishing a rhythm of life without him. This process was considerably complicated by writing a book that was so much a part of him, being committed to doing it as he would have wished when her own way might have differed. When Carol talks of the benefits of taking stock in order to report to Group, I think of this period where she frequently discussed her conflicts between writing the book and moving on emotionally.

Group helped directly with the later stages of the book's preparation. Christine gave substantial editorial input as an expert in one of the areas of research detailed in the book. I agreed to read the entire draft manuscript after Carol told us she had been through it too many times to make clear judgments about content and to reconsider sections that the editors had suggested needed rewriting. I don't know if Carol asked or I offered first, but it was wonderful serendipity for me to take on the task. I was in the early stages of work on this book, and I had the luxury of taking time to read and edit at Carol's convenience. My experience in molecular biology made me a knowledgeable reviewer, but I was not an expert in most of the topic. Because I'd been out of research for almost fifteen years, I had no concerns about who got credit for what. I had remained enthusiastic about explaining biology to young people and nonbiologists, and having taught with Hatch, I felt particularly interested in his approach. Although I was not sure at first that I could help Carol, I discovered that I loved the book's objective and had strong opinions about what worked and what didn't. Carol applauded my suggestions, frequently saying, "I've never liked that bit, but didn't know how to fix it." We rewrote sentences and paragraphs together, enjoying each other's ideas and shared interest in the process of saying it right. What started as an attempt to help Carol out of her writer's doldrums was a tremendous psychological boost for me as I tried to think of myself as a writer.

Portrait sketches of the protagonists drawn from photographs were a particular headache for Carol. At the end of several Group evenings, we gathered around a folder of sketches suggesting changes to the shape of a nose or an eye so that the sketch would more closely resemble the person in the photo. Mimi was the expert here. Her artistic skills, coupled with professional study of form and structure, allowed her to identify necessary changes and communicate them to the artist. She edited all of the hundred portraits. Everyone in Group felt privileged to be with Carol through this project, and we waited anxiously for the major reviews in scientific journals (one negative, unfair, and inaccurate, another glowing and knowledgeable). Each of us was thrilled to get a copy of *Operators and Promoters* for Christmas after its publication in 2001.

In addition to writing, scientists communicate through talks, often at scientific conferences. These meetings can be exhilarating, a chance to learn new things, to see old friends and make new ones. They can also be incredibly daunting, and we've all engaged in mental calisthenics in preparing for an upcoming conference. The greatest pressure is not the talk we're going to give or our concern about our students' presentations but the nonstop, offline interactions. There seems to be limitless opportunity to feel put down, inferior, or simply low in spirit. We devise ways of taking care of ourselves in this context. "Get away from the crowd if it's too much. Have room service instead of going out to another fancy dinner. Get some sleep. If it's a big meeting and your students are attending, decide in advance how much time you want to spend with them introducing them to colleagues and showing them the ropes." Most important, "Leave time for yourself."

Giving a talk is a big deal; one needs to plan, have good audiovisual aids, and practice. Two concerns weigh on the presenter. The first is stage fright, concern for proper delivery and execution, and dealing with the fear of stumbling or forgetting the thread of the presentation. The second has to do with the fact that it is one's own ideas and work. Mimi described this aspect as "trying to be the leading lady in a play you wrote yourself." Public speaking is much easier for some people than for others, and the mood of preparation can range from calm to pure panic, depending on the speaker and the projected audience. For some of us, if we are *not* agitated and worried before a talk, it probably won't be one of our best efforts. The nature of the worry takes a number of forms. Someone said after a successfully presented talk, "I was very worried that no one would come to my talk. My second biggest worry was that lots of people would come to my talk." We have repeatedly given advice about the goal of speaking: "Make it about doing justice to the science, not about selling yourself."

Someone may discuss with Group specific practical issues about an upcoming talk. In my licensing director days, I was invited to speak at a National Resource Council meeting about an aspect of Roche's licensing policy. Certain academic scientists had criticized the policy, so I was prepared for tough comments and questions. Three days before the meeting, I discovered that the folder of background information sent in advance to all participants seriously misstated the facts about the program. How to deal with it? Demand a retraction? Rewrite my talk to address the errors? After Group brainstorming, I went off armed with key advice: "Stick to your original talk. Remember half the audience won't have read the background; if you refer to it at all, do it as an aside." And more generally, "Work at having a good time. Be the observer, gazing down on what's happening. Watch all these people scurrying around. Don't take criticisms of Roche personally." It worked. I gave a good talk, and in addition to the hard questions, some unexpected allies spoke in support of my position.

One experience Group treasures about performance and speaking took place when we attended the cell biology meetings in the winter of 1994. Helen was the only one among us who had never spoken in front of a large audience, and she was petrified. She announced a few days before the session that she had spent months preparing to tell us she *couldn't* do it, but now she was planning to try. As she talked about the persistent fear, we more experienced speakers told her the aim was not to get rid of the butterflies but to get them to fly in formation. We joked that we should have a series of Group merit badges and that Helen was earning her speaking badge. After the panel, as we received compliments, letters, and requests for more information, Helen reported that she was letting in the praise and excitement, and feeling part of it. She told us that in her determination not to speak, to find a way out, she had developed all the reasons why she was separate from Group but ended up realizing she didn't want to be separate. The rest of us, of course, had known she belonged. A PhD and experience in public speaking are not criteria for Group membership.

Returning from a meeting, perhaps having worked on self-preservation techniques beforehand, we enjoy sharing good news about positive feedback we've received. After a talk at an interdisciplinary meeting, Mimi was told, "Yours was the only talk that spanned both groups, and you are the only person who can talk to both groups." We celebrate those triumphs and try to remember them when future crises of confidence arise.

13 *Nobody Taught Us This in School*

INSTITUTIONAL POLITICS AND STRATEGY

- *The rules of the institution cannot be the rules you judge your-self by.*
- *Proceed calmly as if there were no doubt of a successful outcome.*

Many of the problems affecting our professional lives are politi-cal, in the sense that they are about career management or ma-neuvering for power and influence. Although most of us think we would prefer to avoid politics and "just do a good job," invariably we find situations in which our success, or even survival, requires that we understand the political structure of an institution.

"There are no politics as vicious as academic politics, because the stakes are so small." We often quote this witticism in that it captures the absurdity of many hotly contested issues and casts a humorous light on some of our struggles. But even absurd problems may have a significant effect on our professional lives, so Group members help one another decide when it is safe to ignore a particular issue and then address what can't be ignored. Most of my examples are drawn from academia. My experience there culminated in my departure from the university, yet the academics in Group who did get tenure had strikingly similar experiences—a sense of powerlessness, overwhelming workloads, and competition for access to grant money, lab space, and students. The same concerns arise in other professional environments, although the details may differ.

A common theme of our work is that large impersonal institutions are just that—impersonal—and we should not take every negative situation personally. We discourage the tendency to assume a victim role and remind one another that there is no "they" out there acting with the objective of bringing us down. Almost everyone has been disappointed when people fail to follow through on assurances of concern or intended action. As an illustration, a member of Group was being courted by another university, and for months she got no indication that her current institution cared what she did. Then suddenly there was a flurry of activity. "I got messages from both the dean and the department chair saying the chancellor was trying to reach me and was extremely concerned that I might leave. . . . Then I met the chancellor and it was clear he didn't recognize my name." Group reminded her that although people seemed insincere, saying only what they thought they should say, it was unlikely that they consciously chose to lie to her. Indeed, the chancellor probably was upset to learn that the campus might lose an outstanding faculty member but wasn't close enough to the situation to know her by name. We laughed a little, sympathized a lot, and were pleased when she decided to stay, since the institution that was wooing her was out of Group range.

Sometimes a good report to Group helps shake off the Victim Pig when it seems like the world is dumping on us. Mimi shared a story that begged the interpretation that "they" were conspiring to make life difficult. She had recently turned down a job offer from Harvard after a long and difficult decision-making process, focusing instead on moving into new lab space at Berkeley. "The floors needed to be sealed, and I requested that it be done before the furniture was moved in. They said no; it was no problem to do the floor work around the furniture. On the day the sealing job was scheduled, we shut down all our experiments and left the lab as we had been instructed, ready for the work. The janitor came in, looked around, and refused to seal the floors because the furniture was there. I lost it; I jumped up and down screaming at the poor janitor. Then I sent e-mail to other faculty in the department entitled 'Why I wish I'd gone to Harvard.' And then I went home and threw up." In a way Mimi was experiencing the inevitable roller coaster of ambivalence that frequently follows a big choice—but she let it get to her and had regrets about the e-mail. The incident brought all her frustrations with Berkeley to the surface, and although she knew that carelessness rather than malevolence was the cause, she had to struggle not to feel victimized.

Sometimes an action of a colleague *is* directed at us. Yet the behavior may not be about anything we have done or can control but simply that person's way of exerting power. When someone reports being put down or manipulated by a peer or a superior, Group can be useful in identifying a power play. We are usually best served by stepping back, not taking the interaction personally, and proceeding calmly. I received valuable advice in this arena when I was dealing with the only bad boss of my career, a man who seemed to enjoy belittling me at every opportunity. Group pointed out that I had control over how much power I gave him. "The worst thing you can do to him is to be fully professional. Distance yourself from him, don't try to like him or have him like you, and it will get easier to lob back his volleys." I learned to ask questions. "You asked me to do this job. What exactly is it that you wanted to have done that I have not done?" (He had reassigned a major area of my responsibility while I was on vacation.) After he'd given me an unexpected directive and I hadn't been able to think it through on the spot, I responded, "There were a number of things that weren't clear; I need more information." When I followed this tack, doggedly returning to the professional issues and ignoring the personal barbs and sudden shifts, it became much easier to deal with him, as Group had predicted. A general lesson for this type of situation is to consider whether the source of an attack is someone who may feel threatened by the competence of others.

Another event that affects our professional lives and that many of us have struggled not to take personally is the departure of a treasured colleague. One fights the certainty that things are going to fall apart. This is especially tough when the departure is motivated by disapproval of the direction the organization is taking, which invites the question "Am *I* crazy to be staying?" Even when someone leaves without rancor to join a spouse or to accept a fabulous new opportunity, those left behind may feel abandoned. Suzanne reported grief and fear for her institution's future because a prized colleague had taken a job elsewhere. "Will the place fall apart?" she wondered. Judith assured Suzanne that her response was natural. "I feel it whenever a very good postdoc leaves the lab, even though it is part of their normal career progression." Christine told how people felt after the sudden tragic death of the beloved professor who had built her department and recruited most of the young faculty. "There was total panic. We were sure everyone would leave. Then Bruce [the replacement chair] came, and it was fine." Suzanne then asked for specific advice, because it would be her job to recruit new people, a task she resented since she would have preferred that her colleague stay.

Group advised that she take on the assignment with enthusiasm, even though she grieved the reason for it. "Consider it a chance to shine, to influence the future."

When we have little control over events, as in the examples above, we work on responding appropriately and taking positive action. Equally often, we are presented with choices that are clearly ours to make but are complicated. We seek one another's advice about individual responsibility to the scientific community and to the institution that employs us. We weigh carefully whether to take on service roles that will distract us from primary professional objectives. At the university, for example, numerous faculty committees often demand inordinate amounts of time away from research, publishing, and training students. We have encouraged one another to serve on those that seem most rewarding and to exert our power to refuse those that are not. Carol poured enormous energy into an outreach committee dedicated to increasing graduate student diversity at the university and to supporting minority students already there. She cared about this commitment so deeply that it was worth the drain on her time. Deciding is always a question of balance, asking whether a service task makes sense, what the ramifications of refusing it are, and how much time and energy it will take away from our individual goals.

Senior university faculty members are often asked to take on high-level administrative responsibilities. Those who do accept often have misgivings as they assume these roles, struggling with how being a dean or provost or department chair will take attention and time away from science. The motivation to take on such a function is concern for the health of the institution. Faculty who depend on an institution for professional well-being feel an obligation to give something back and may be committed to certain improvements. When Judith agreed to be chair of the Department of Chemistry, for example, she did so with specific objectives that made the burden more acceptable. She stated her commitment to increasing the importance of biology in the department and to promoting diversity in new faculty hires. (In the first of these she was very successful; the latter proved much more difficult to achieve.)

There are numerous parallels between my experiences in the corporate world and those of Group members who take on management responsibilities outside their labs. Some have reported unexpected pleasure in the problem-solving teamwork involved in administrative work and have noted that administration tends to include more tasks with a clear end point than research affords. This can be a refreshing change. Group work for those taking on substantial efforts outside the lab is often directed at finding ways of remaining in touch with the research staff and scientific programs. Solutions have included strict adherence to schedules that restrict certain days or hours to science. One Group member, relieved to step down and return to her lab after serving a long term in administration, soon found that she missed the interactions and her participation in efforts to make the university an effective and hospitable place to do research. "Since I left [administration] I feel like I'm staying home with the kids," she said as she focused on the narrower concerns of her own lab. Although individual members of Group each have their own reactions to these decisions and are always protective ("Oh my god, are you *sure* you want to do that?"), we respect each other's choices and participate in the inevitable struggles of increased workload and competing demands that follow acceptance of such a position.

Obligations to the scientific community outside the institution include serving on the editorial boards of journals or the study sections of funding agencies. Study sections are committees of scientists established by funding agencies to review scientific proposals and recommend which ones should be funded. Participation is time-consuming, but the system of peer review of research and funding requires that people contribute their time. All the Group members who are research scientists have served on these committees. Suzanne, having agreed to this outside commitment, brought up the issue of controlling the amount of time she spent. "I need to learn to review grants without rewriting every page, to keep the editor in me from taking over." The reviewer's job is to comment on strengths and weaknesses of the science proposed, not to edit the presentation. She also noted that she was writing her review of each grant request as though it was a scientific paper of her own, so the writing took an inordinately long time. Group suggested an affirmation: "The idea is not to create a perfect review. It doesn't have to be a masterpiece." Suzanne made a contract to streamline her reviewing process. One technique was to limit the amount of time dedicated to the task by starting later (closer to the deadline). We reflected that it is tough to strike the right balance. Because a reviewer knows from experience how much effort goes into preparing a grant request (and may have had the discouraging experience of being reviewed carelessly), she wants to show respect by doing the job well. That attitude is essential to the fairness of the process, but to put many hours into graceful phrasing of the written report is a waste of time. This is a situation in which Group members' feedback is truly helpful in deciding "how much is reasonable and how much is too much."

One condition that is essentially political is preparing for situations in which we know we are likely to feel intimidated. Group members' contracts often incorporate the concept "Act as if" to address anxiety: "Act as if you were completely confident." "Act as if you had done this before." We use the safe haven of meetings to explore our doubt and insecurity, picture how it would look and feel to be confident, and then try to proceed in that way. One example was preparing for my job interview at Cetus. At age thirty-six I was seeking a nonresearch job for the first time since

I was seventeen. Everyone agreed that a state of high anxiety was perfectly natural and suggested a contract we had used before: "Proceed as if there were no doubt that you will succeed." Variations on this contract continued to help me after I got the job and had to adapt to a new environment.

University life can be crazy for anyone, but being female has figured significantly in our experiences. As individuals and in our meetings we have been determined to avoid the habit of assuming that every setback or problem is gender-related, but when it is, we have to recognize and deal with it in that context. In the early years, many of us shared the experience of being one of few women, or perhaps the first, in the role we had taken on. As we entered environments that were predominantly male, even the most enlightened of the departments (I claim mine as the least enlightened) had faculty members who were at a loss when dealing with a woman as a professional equal. Administratively, we might either find ourselves asked to be on too many committees because "we need a woman" or left out of the loop on some topic because there was no obvious women's issue.

One feminine trait that we have identified as a deterrent to success is the instinct to be polite at the expense of effectively making a necessary point or challenging someone. I have often exhibited a tendency to choose nice over forceful, and Group lets me know when they perceive that it is lessening my effectiveness. As I garnered professional experience, I gradually got it that men were given much more latitude to be opinionated, out-spoken, and aggressive. Not all men act that way, of course, but those who do are not penalized as women are for the same beha-vior. Since in general women *need* to fight harder to get the at-tention they merit, there is a vicious cycle in which women are disliked or discounted if they exhibit strength but fade into the woodwork if they are too docile. In our Group work, this dilemma is reflected in contracts that include a resolution to be firm and to speak clearly to our interests, even at the risk of appearing aggressive.

Judith's experiences with gender bias are especially absorbing, in part because as an assistant professor I so envied her security in having come to Berkeley with tenure. I was awed by her stature and assumed that she was immune to the sense of isolation and embattlement that I felt as a junior faculty member. In fact, even she found herself marginalized in ways remarkably similar to my experience. It is as though men in positions of power saw hiring a woman as a one-step achievement, not considering that there-after they should consult with her, honor her opinion, and treat her as they would a senior male colleague. In one situation she reported feeling helpless and politically impotent because no one had sought her opinion on an issue she cared about. She wondered whether it had to do with being a woman. "I don't know how to get what I want, and if I can't figure it out, the de-partment may go in a direction that is not what I think best, with-

out my input. I don't know how to proceed in my own best interests." Group advised, "Be clear with yourself about what you
want to achieve, so that you can communicate it." Preparing for
the meeting at which she would address the issue, we urged, "Establish your commitment to this process. Don't be polite."

Several of us have been drawn gradually but steadily to an increasingly feminist perspective by situations for which gender
bias is the inescapable explanation. We have learned to look out
for younger colleagues and to be sensitive to inequities in evaluation and promotion. I look at it as maintaining a belief in the
possibility of a level playing field without getting lulled into
thinking it has been established. In my years of overseeing employee evaluations, I learned how easily a boss could criticize
style when he (or she) honestly thought he was criticizing performance.

As our careers progress, we hope that we have gone beyond
personal experiences of discrimination. Alas, it is not wise to get
complacent. As a full professor, Judith discovered one of the
most demonstrable forms of gender bias directly affecting her.
She first told Group that she believed, based on information she
had obtained with assistance from a colleague, that men in her
department ten years less senior than she were paid substantially
more. Through a review of publicly available records, she found
the situation was worse than she had suspected. The discrepancy
had been initiated at the time she was hired, and the salary gap

had widened over the subsequent fifteen years. She had wrongly assumed that administrative oversight of salaries and pay scales assured equity, and the discovery left her overwhelmed by anger and a sense of betrayal. This revelation was all the more critical because low pay in the present would affect retirement pay in the future. At a time when her substantial scientific achievements were being recognized by her election to the National Academy of Sciences and by numerous other honors, she was faced with evidence of blatant discrimination.

Judith fought for herself by asking for a raise (which she received) and back pay (which was denied) and by consulting a lawyer. Despite attractive job offers from other universities, she wanted to stay at Berkeley. Knowing how shabbily she had been treated would make that difficult, but she didn't want the situation to drive her from her community. Also, she realized that the discrimination was unlikely to be limited to her. She told Group, "I realize how deeply this has affected me. I have previously denied that I cared very much." She could not let go of the issue, however, for herself or others. As a plus, she recognized that she could "relish the camaraderie with female colleagues." Ultimately, the university undertook an investigation of salaries in her own and other departments with the goal of addressing gender inequity.

Struggling to figure out how to survive as political animals in an institution has fueled our desire to help others in the same struggle. While solving problems and supporting one another remain Group's principal objectives, we have become more and more interested in sharing the group experience with others. We have been concerned that Group is exclusionary because we aren't open to additional members, yet we are aware that increasing membership would imperil our own work. So we have tried to share our collective wisdom by giving talks, both individually or as a group, and by having one-on-one discussions with junior colleagues. Invariably we find that both the general and the gender-related issues that have absorbed our attention are common to others. The concept of regular meetings, with the objective of advice and support in a confidential context, is critical to groups and other mentoring relationships. To be able to share "It makes me crazy" can be a tremendous relief, but the practical discussion of "What do we do next?" is even more important.

A few weeks after our 1994 panel discussion, a postdoctoral fellow at Stanford who had been in the audience visited Berkeley with a colleague. Their purpose was to meet with Christine and me to discuss starting a group of their own. It was a great pleasure to receive a message a year later:

> Just a quick note to let you know that our problem-solving group is well and alive. We celebrated our one-year anniversary in February and we continue to grow! There are eight of us in total and we call ourselves the WET ones,

for Women of Exceptional Talent. We are thinking about an overnight retreat in the next few months.

And five years later:

Happy New Year!! I was thinking of you yesterday as our problem-solving group met at my house. We were remarking how amazing it is that we have held together for six years, in spite of the fact that we are now scattered all over the Bay Area! It has evolved into a wonderful community of women scientists and provides a meaningful connection for me, especially in the midst of our rushed lifestyles. Thank you for planting the seed at the ASCB!

This is one of several groups that have been inspired by our model. We are particularly interested in communicating with women, because of our awareness that women often lack mentors and training in political skills, but we are delighted when our message appeals to men as well. Whatever the audience, our advice focuses on the importance of talking out not only problems but also procedures and progress toward goals.

We recommend groups from our own experience, but a group can't do everything. In most professional environments, one benefits from having a mentor or advocate within the organization provide perspective and advice. The importance of creating and maintaining a relationship with a mentor is part of our mes-

анонanc оригинаurst

sage. Often a Group member reports exultantly that someone she has counseled has contacted her to say that things are better, that she has established a mentoring relationship, or that she has gotten together a group of colleagues to discuss common problems. We celebrate these stories as personal achievements and as evidence of Group's principles having an impact beyond our immediate ranks. Although I don't think any of us believes that the toxic aspects of institutional politics will soon be eradicated, people committed to open discussion and the airing of concerns will make it easier to survive them—and will nudge the institutions toward change.

14 *Anticipating Changes*

GROWING OLDER WITH GRACE

- *As I get older, there are things that are driving me crazy.*
- *It is not so much that I feel wiser; it's that I accept things more.*
- *It's hard to get the concept of aging. Sometimes it seems clear; sometimes it's a mystery.*
- *I'm not going to be able to climb down bluffs on a rope much longer.*

The adventure of aging has become increasingly absorbing to Group as we anticipate this later stage of our lives. We find our attitudes changing, sometimes in unexpected directions. Taking the time to figure out what we really want becomes more important as diminishing time and energy become more acute.

As the eldest, Helen is often our role model. One evening before her sixtieth birthday she told us, "Sometimes aging hurts and I cry a lot. Sometimes it feels fine. The bad feeling doesn't stay long." The rest of us, all under fifty at that time, took note, as we did when Helen retired. Her process started with acknowledging that she was unhappy with her job. Looking for a new one, Helen listed the things she liked about her work: "communicating information, making decisions as part of a team, allowing others to participate, and working with smart people who have a sense of humor." This exercise brought into focus some good things her current workplace provided and what she would seek in a new job. Ultimately, she chose to retire from the university instead of finding a new job there. Her first report from retirement was positive. "I'm doing things when I want to. It is such a good feeling to have done this for myself and for others in my life. I don't feel exhausted, working part-time." Not feeling exhausted was a delicious fantasy for the rest of us.

Helen also tried some new things. Describing Latin dance lessons with a friend her age, she commented, "As you get older, you can be more outrageous." She told us about one of *her* role models, her friend May, "the one who taught me irreverence when I was twenty." May, at eighty-six, had been through a bout with cancer, was suffering from glaucoma, and could no longer drive, but Helen still treasured May's irreverence and hoped she retained that image of herself.

We have shared social, professional, emotional, and physical experiences of aging, each of us with unique variations on common themes. Judith returned from a trek in Nepal, having found that, not yet fifty, she was the oldest person on the trip. It was not unpleasant to be the senior trekker, but it was a surprise, a gentle prelude to being more and more frequently an "elder" at a university meeting or professional conference. There have nonetheless been some bumps in her acceptance of seniority. When Helen mentioned that she was enjoying the benefits of the American Association for Retired Persons (AARP), Judith exclaimed, "Oh! They sent me an application. I threw it out. I couldn't bear it!"

As we grow older, we need to slow the pace of our lives. This, in turn, is inextricably tied to considering our desires. If we keep too busy, as we often do in our younger years, there is no time to notice what we want. My decision to leave my corporate job to let new things into my life followed many years of thinking that something was missing. Eventually I gave myself time and mental space to consider what it was. Of course, quitting one's job is not always a viable option or the coveted fantasy that it turned out to be for me. Sometimes the idea of quitting is an extreme proposal from which to consider a less drastic response, as when Suzanne said, "I think I have to retire," paused and added, "or at least cut back on working weekends." Most Group members have worked on carving out time for their nonwork selves *without* relinquishing cherished career goals. Whether the goal is to spend more time with a child or in the garden or exercising, we work to readjust our schedules and sense of obligation to make it possible. The struggle to make time for what enriches us is not necessarily a battle between home and job; it may be about how we live life at work. Mimi once said, "I have nice colleagues and I have no time for them. The lab is really wonderful and I have no time to spend there." These considerations have become more important as we get older.

Professionally, our status and power are altered as we age, as are our attitudes toward our work and the people around us. We notice a dichotomy in our capacity to be patient: we seem to gain patience, because we can tune out bullshit, yet we are moved to *im*patience because we feel our time is limited and we want to conserve it for important things.

Some of our experiences of aging are intensified through watching our younger colleagues. A mere four years after getting tenure, Christine reported, "I went to a Cold Spring Harbor meeting and my students talked instead of me. I'm feeling like Old Mother." Teaching a summer course, Suzanne loved meeting the excited young students, whom she found "hot and smart," but didn't like their making her feel old. Judith gave her feedback: "Remember this is a new time in your life. It's not the same time as being a postdoc or a grad student. You have other interests." A few years after that exchange, Judith talked of having too little time left before retirement to do the experiments that intrigued her. "I keep thinking, 'That's a great idea but I'm too old to start.

It's too bad I became so successful at this age.' I'm fighting a lack of energy." Christine suggested that it was time "to exchange energy for clarity." Sometimes it feels overwhelming to be losing energy and vigor. Acceptance that aging is at times (but not always!) sad and frightening has been a take-home lesson of our discussions. Because we generally feel that our jobs demand more than anyone could possibly accomplish, even with youthful stamina, a diminution of our accustomed energy is quite terrifying.

We have tried to be aware of and attentive to, but not obsessed with, the aging of our bodies. As we moved through our forties and fifties, Group work began to include reports of hot flashes, pulled muscles, discussion of estrogen therapies, and the chronic exhaustion that is so different from the healthy tiredness that a good night's sleep can cure. As Christine put it, "Being forty-five is like being an adolescent: there are so many changes, that you can't ignore your body."

We've reaffirmed the need to find inspiration in others' ways of dealing with illness, aging, and death. As Mimi said, "Genetics suggests I may die in my seventies as my parents did. I need to live, not just to survive." She made a contract to "notice and celebrate the time while I am alive." Our own illnesses, from the routine to the terrifying, are juxtaposed with the experiences of those around us. Helen's brother and my brother-in-law were both diagnosed with terminal cancer the same summer that Suzanne had severe peritonitis that made us fear for her life. Suzanne's experience brought back Mimi's memories of a time Mimi herself had been hospitalized and in danger of dying, and she spoke of how tenuous life seemed. Later Mimi and Suzanne, comparing notes of their experiences, concluded that "life is very sweet when you have been seriously ill." Helen echoed this thought with a quote from a favorite Buddhist writer: "The more you feel impermanence, the more precious every moment is."[1]

One of our most fearful reminders of mortality followed Christine's discovery that she had breast cancer. After her surgery, we were initially elated when her doctors reported that she did not need chemotherapy and then agonized over the revised advice that she did. Her description of the feelings associated with her treatment and diagnosis was clear, inspiring, and scary. "There is a new personal reality that comes from having cancer. When someone says you have a 30 percent instead of 20 percent chance of metastasis, what does that mean in life terms? Do you do something differently?" During Christine's chemotherapy I found my mind imagining the worst possible outcome, as I often do when people I care about are ill. This made me feel disloyal, as if my failure to be purely optimistic was a failure to support Christine. When she said in Group, "It's hard to think about the future with me in it," I cried, finding it too painful to think of the future *without* Christine in it.

We've worked, too, on the practical details of growing older, writing wills and trusts, embarking on new phases of financial planning, thinking about the possibilities for retirement living. Helen reported on the death of a friend and made a contract "to move the Living Will book to the top of the pile." This led to much discussion and a Group contract to help others with these issues; we chose to look at them as milestones, just as buying a house, getting a job, or having a child finish high school are milestones. When Helen completed a living trust and power of attorney, she said, "It feels good to have done it, confronting impermanence." She contrasted the present with her past: "I'm getting stronger. I remember twenty-four years ago, buying the house alone [after her divorce] and getting a small loan for landscaping. I think of how I've changed, how the status of women has changed."

As we continue through the process of aging, altering the pace, dealing with physical changes, and weighing retirement against ambitions, we try to trust our instincts and remember what's important to us. Helen developed an affirmation that described the ideal as "flowing with grace into the future with my dreams and self-trust." We recognize the importance of growing older in the company of people we care about. Group provides us the opportunity to do just that, comparing notes along the way.

15 *Going Home*

- *I want to love my family, to value their love and respect, but to define myself.*
- *The more I let go of my daughter and let her be who she is, the more I let go of my mother.*

Group discussions often concern spouses and families, both their impact on our professional lives and our preoccupation with their problems or their responses to ours. As the years have passed, our discussions about family have become more personal. The spouse who *still* hasn't seen a doctor, a child who is floundering, and a mother with hurt feelings are familiar subjects. Just as with professional issues, we have been able to draw on others' experiences to get new viewpoints as well as practical solutions.

When we talk about family concerns and frustrations, we demonstrate a level of trust beyond our formal rules of confidentiality. By considering it safe to discuss my relationship with my husband, I'm trusting Group to give feedback, and probably sympathy, without making judgments. To bring negative issues about family members to Group, we have to feel we're not being disloyal, whether we are in major emotional turmoil or just trying to get a partner to take out the garbage. This is especially important, because we interact socially outside Group, often as couples.

We have identified various complications in getting family support on professional issues. Although conflicts between commitments at home and commitments at work certainly exist, they have not been the only focus of these discussions. A spouse may have the best intentions and be fully supportive of his or her partner's career, yet not know how to provide encouragement in a way that will be perceived as supportive. Group often can discern the gaps in understanding when someone describes disappointment in a partner's response, which sometimes results from expecting our loved ones to understand how we feel without our telling them—a tall order.

Another gap between reality and expectation is that when we seek comfort, a partner may feel required to do something. The concept of "just comfort" doesn't come naturally to people used to action, which includes members of Group as well as our partners. If there's trouble, our instinct is to try to fix it. A recital of complaints, fears, or failures precipitates a "need to help" response, when maybe all that was wanted was an arm around the shoulder. Christine said her husband got defensive when she talked about a problem that he couldn't solve, as though he felt he was failing her. Group suggested that she not try to change his response. "When you find you are asking for comfort and are feeling unheard, just recognize it and take care of yourself." We have incorporated this understanding into Group process. When the person about to work states what she wants from Group, "just comfort, please" is an alternative request to our usual directed problem solving.

A related idea that has been useful to me is that there are times when a partner *can't* be a comforter. When I was struggling as an assistant professor, I told Group that I wasn't getting the support I wanted at home for my anxieties about my department and my lab. I felt I was rehashing the same fears over and over, and David had nothing to say. The suggestion from Group was that I use David as a refuge, save him as a place *not* to deal with my most frightening feelings about work. They pointed out that seeing me in a situation of despair was hard for David. Hopelessness is repetitive, and it is painful to hear the repeated terrors of someone you love when you have no power to fight them. It's not a question of shielding a partner from the bad feelings; it's realizing that home may be more comforting if we don't demand to talk about our fears there. "Use Group for that" is the advice.

Group is committed to listening, criticizing, and suggesting without taking on the situation. Responsibility for finding an answer and choosing a path remains, as it should, with the one who owns the problem.

I think I make fewer impossible demands on my marriage because I have the opportunity to discuss my thoughts with Group. When I learned that I wasn't being recommended for tenure, I was grateful to David for his rage on my behalf. I cried on his shoulder and loved the roses he brought to cheer me up. But I did not completely welcome his confidence that I could turn the situation around. When he said, "Just wait to see the flowers I bring when the decision is reversed," I felt pressure to perform in a certain way, to fight the decision even if I didn't want to. I appreciated the security of my marriage, but at the same time I resented that I wasn't on my own. Alone I could have behaved badly, burned bridges, run away, had that nervous breakdown, left town, all of which seemed appropriate and attractive. Being married appeared to rule out those responses. I had to act responsibly because of obligations to another person. I used the "refuge" concept Group had suggested and focused on the way my home life helped me keep my balance. When the time came to make choices, I felt glad to have someone else's needs to consider rather than feel tied down by family. I could justify not considering jobs outside the Bay Area because my stepdaughter was in high school there and David had a job he loved.

Several of us have worked in the same field, or at the same institution, with a partner. The upside of this situation is the ability to share, talk, or work through problems together and have
our issues thoroughly understood. David and I found it difficult
when both of us got angry or depressed about the same events at
work. On the other hand, if we had opposite reactions to the
same situation, the one who was pleased resented being brought
down by the dissatisfaction of the other, who then felt unsupported. When we were both at Cetus, I complained that I was
feeling harassed by his anger over low salary increases available
for people who reported to him. Salary increases were a major
headache for me as head of personnel, so I hated talking about
them at home. Group had unanimous advice. "This is a conflict
of interest. You can't be part of the discussion. Tell yourself, 'I
have a right not to talk about this now' and stick with it." Then we
slipped into silliness. "Buy some champagne and get him into
bed. Every time he mentions something about Cetus, remove
another article of clothing." I never carried that plan out, but
David and I did resolve our discussions of contentious work issues and for the most part have each enjoyed having the other as
a sounding board.

Having a partner with a different career focus and different interests has its own set of challenges. Some Group members have
reported having their efforts at science and university work discounted by partners whose focus was on artistic, spiritual, or psy-

chological pursuits. These differences are especially tough because such questioning inevitably stirs up self-doubts. The level of effort we put into academic scientific jobs does indeed appear insane at times, and it's a struggle to distinguish the necessary from the excessive. Their reports were a thought-provoking contrast to my sense that David would be disappointed in me if I *didn't* work harder and achieve more. Suzanne's experience was a bit like mine, finding sometimes that it was burdensome that her husband, Chris, put such a high value on her scientific success. Hearing one another's reactions gives us perspective on our own.

Partners in a close relationship neglect each other sometimes, whether inadvertently or because of a need to focus on a problem that may not directly involve the other. Most of us have been on both the giving and the receiving end of that kind of neglect and have been confronted with a pouting spouse or hurt by apparent disregard. The bottom line on getting and giving support on professional issues seems not so different for those who work closely with our partners as for those who don't. The key is to recognize when the other needs support and when we do ourselves, and if the two conflict (as in both being needy at the same time), to try to find a balance. One recommended affirmation was to say, "I love you, I love myself, I need distance right now." The hard part is to get rid of the guilt at not solving someone else's problems or taking on their pain. "When someone is sick, you take them chicken soup, but you don't have to catch the cold."

We've traded suggestions on how to argue, discuss, and ask for what we want. One rule that is sometimes difficult to follow is to stick to the present in a conflict and not to let the past take over. Suzanne worked on this chronic problem over several years, saying that she hauled stuff out of the past when she was angry with her husband. "It comes out as 'Fifteen years ago you said . . .' and I invalidate my present feelings by dredging up the past." Group suggested that she write down past issues when in a calm state. "Deal with the old things, but don't apply them to today." Ultimately, this worked. Suzanne told us that she realized she was no longer angry with him about things that had happened previously. "I believe Chris is on my side."

Shared stories of interactions with our families—spouses, siblings, children, and parents—have provided useful perspective. Judith told us that on returning from a trip she has a compulsion to water plants and to fix everything in her house before relaxing. This irritated her partner, who wanted to welcome her home with a glass of brandy. Her story shed a new light on my own reactions. I had always hated David's insistence on unpacking, doing laundry, and sorting the mail before relaxing after a trip. Now I think of Judith whenever David does his "coming home" routine, and instead of pouting I help with the laundry and the mail so we can get to the relaxation.

We've dealt with fears about the health of our partners, which are sometimes harder to deal with than our own health issues. Many of us have struggled over fear of nagging. How do you respect the other person's right to do things his or her way yet tell them what you think about their weight, eating habits, or refusal to see a physician for that lingering flu or swollen leg? The positive response most of our families have toward Group suggests that they think the advice we get there works, for us and for them. Family members sometimes note that we seem to feel better after a Group meeting, thereby acknowledging the usefulness of a forum for discussion outside the home.

Children are another family topic. Someone may be concerned about a child's well-being and doubt her ability to provide guidance. With older children, we seek advice on staying out of the way (when that's warranted) and on practicing the art of letting go. Our members represent a range of mothering roles, including the traditional two-parent variety, single parenting, and stepmotherhood, each with its own issues. When Helen asked for feedback on how to deal with anxieties about her daughter (then in her early twenties), Group suggested that parents tend to interface between kids and reality. "It is as though there's an agreement that if they tell you all that's wrong, you will make it right. We have to let go of that." Surprise! This falls right in line with our work on a partner's assumption that a statement of unhappiness includes the expectation of having it fixed. As children grow up, they may appreciate, as much as we do, the opportunity to talk out problems and expect comfort, though not a solution. We need to encourage that part of maturing.

Our emotions about the young people in our lives range from worry to love, frustration to admiration. Judith, as a divorced single mother in a highly competitive career, has worked on many classic conflicts about time spent with her sons versus her commitment to work. When I hear her speak of her younger son, now married and father of two, I remember with relish her statement from the mid-1980s: "My son is one of the most wonderful little persons that ever lived." This expression of joy is particularly wonderful because it was the bright spot in a painful session about loneliness and doubtfulness of her performance as a parent. Suzanne, who also struggled to balance career and family, reported, "I like Maria—even when she yells at me. I would not be willing to be a scientist if I had to give up my daughter." She went on to say that, having arranged to spend time at home with Maria over a holiday, she was appreciating her mothering and home-making sides. As a bonus, her reallocation of time had forced her to set priorities, with the result that she actually got more done at work.

The stepmothers among us have shared experiences of sometimes feeling excluded, sometimes close. We've gone into self-protective mode, not wanting to initiate contact lest we be hurt by rejection. Other times we have let go of caution and accepted vulnerability, which may have great rewards. Bringing to Group some of the sorrows and trials that we stepmoms assumed to be unique to our situation, we discovered that the experiences of real biological mothers are not so different.

Group gave me lots of support in a crisis with my stepdaughter Duskie just before she started college. She wanted to leave in the middle of a family vacation to spend time with high school friends and approached me instead of her father. I mediated to arrange her departure, then regretted it. Because she had never explained herself to him, he was hurt and angry, though he didn't tell her so. I felt my interference had kept them from working things out. Group's first insight was that "they might have done more damage to their relationship if they had talked." Defensive feelings might have made a discussion impossible. The second point was that I craved a direct relationship with Duskie, and I needed to tell her that I wanted to be more than "Dad's wife." Finally, Group pointed out that in every family someone takes the role of the fixer, and I was doing just that. I could learn to stop taking on the responsibility for family harmony.

I made progress at relinquishing the fixer role and was rewarded by seeing David and Duskie's relationship mend and flourish without my interference or assistance. Following the advice that I pursue an independent relationship with Duskie was painful in the near term. I told her in a letter that I wished to do that, and she responded politely but firmly that she didn't have time. I shed many tears as I worked to see from her side the burdens of having a mother, a father, stepparents, stepsiblings, and a college life to live. Group joined my laughter as I recognized that while we *struggled* with contracts to be totally honest, to ask for 100 percent of what we want, and to set limits when too much was asked, Duskie was already there! She had heard my request and answered with what was right for her. I should use her as a role model instead of wanting to curl up and cry.

Over time, Duskie and I developed a warm, rich relationship that is very precious to me. Others, too, have developed mutual appreciation and adult friendships with a spouse's children, despite complicated early histories. One member of Group reported that her husband's daughter brought up the subject as an adult, saying, "I must have been really difficult for you to deal with when I was a teenager and you and Dad were living together."

Some of us are now grandmothers, or stepgrandmothers, with a whole new set of relationships to discuss. Before Christine and John were married, she told us she didn't want his daughter's children to call her "Grandpa's girlfriend." We brainstormed possibilities and Mimi came up with "Significant Grandother," which captures the ambiguity and delight of the complex modern family. Among the joys of this stage is the opportunity to see our children as parents and to admire their energy and capacity for love. Some of us of course encounter the challenge of letting them do it their way, whether the "it" is child rearing, house buying, or pursuing a career.

While we work on our mother and grandmother roles, our relationships with our own mothers continue to engage us. (We seem to talk less frequently about fathers, although several of us have had warm, important relationships or conflicts with our fathers.) There are two principal themes of our discussions about our mothers. One focuses on the past, as we try to better understand them in order to understand ourselves. We search our mothers' lives and characters for traits and choices we'd like to emulate and others we'd like to avoid. Someone described her complaining mother as "the original victim." In affirming that I had never heard my mother express regrets about major life choices, I recognized a source of my own security. Mimi recognized some good in a deeply destructive relationship with her mother when she told us, "She never liked anything of what I was . . . but we shared a love of thunderstorms."

The second theme is dealing with our mothers in the present—their needs, and sometimes demands, and our responsibilities and commitment to them. One of us characterized her mother as "spunky, intelligent, neat, and a pain in the ass." Another, accepting her mother's love, said, "I know I'm the light of her life—it's been a burden, now it's a pleasure." And a third recognized her mother's expectation that her daughter supply things that her own life had lacked: "My mother is sad, but I have to be happy for her."

The individual daughter-mother relationships of Group members span many degrees of love, trust, friendship, and conflict. Some of us are enriched emotionally by our mothers, others drained, and some may feel both ways at different times. The common thread is that each of us has been profoundly influenced by her mother and has tried to come to terms with that maternal power.

Each of us has compared herself to her mother, either in apprehension of repeating a pattern we don't like or in recognition of a strength we admire and hope to emulate. I have done many things differently from my mother, from studying science to marrying late and bearing no children, yet I am very like her emotionally. Some of us have dedicated considerable energy to being different from mother. This may be constructive, if we see a way that we want to be different in the world. It can also be draining and counterproductive, if we fight mother's real or inferred disapproval as we make decisions about our lives. Sometimes it's useless. After all, genetics and upbringing determine who we are, and a mother contributes to both.

"I can be different from my mother without condemning her," Helen commented a year after her mother's death. Following a path that diverges from that of one's mother, as we all did, can complicate understanding and being understood by her. Helen noted that if a daughter makes new and independent life choices, she may appear to devalue her mother's, and a mother's criticism may be a fight against perceived rejection. In return, we as daughters have talked of feeling guilty at having more of something than our mother has, be it material possessions or success in the world outside the home. Just as we affirm that changing direction does not invalidate previous choices, we remind ourselves that following a path different from our mother's doesn't mean we think hers was bad.

Group saw Helen, Mimi, and Christine through the deaths of their mothers, sharing their grief and foreshadowing their own inevitable loss. Mimi observed that "when mother is gone, we still feel the need to criticize ourselves if we stray from what she thought right." She commented that her own self-image was so deeply rooted in rebelling against her mother that she had to re-examine her life when her mother was no longer there. She pondered that perhaps she had become a scientist instead of an artist so she could hurt her mother. If that were true, she said, "now that she's dead do I still need to be a scientist?" Amazing to hear from one of the most talented and passionate scientists I have known that she'd taken that path in rebellion against a mother who cruelly and consistently put her down for her looks, behavior, and choices. In opening up to consider the whole picture of our pasts, we may also continue to discover and remember the things we shared with our mothers and learned from them, as

Mimi did reflecting on her mother's love of thunderstorms and her knowledge of art and ballet.

Incidents in our own mothering roles may elicit memories of our mothers. Someone spoke of her concerns about her son, who did not have a girlfriend and was upset and lonely. I responded by revealing that when I was a chubby, unpopular teenager, my mother always said, "Don't worry, dear. You'll be attractive later" or "Don't worry that you are tall, the boys will catch up." Christine, the tallest of us all, whose teen years were publicly chronicled in her mother's popular humorous books, cracked, "*My* mother never said not to worry." Mimi added, "*My* mother said, 'Why aren't you *more* worried?'" We dissolved in laughter, and the concerned mom of the present got some insight on ways to deal (or not deal) with her child's heartache.

Judith captured one of the most important lessons about filial relationships when she said, "I must let go of the things that mother can't give me and appreciate the things that she can and does give." It strikes me that while striving to achieve this, we might also apply it to our other loved ones and hope that our daughters, sons, and partners will be able to reach the same place with regard to us.

Part 3 A Group of One's Own

16 *Pigs, Contracts, and Strokes*

GROUP PROCESS AND HISTORY

- *That sounds like your pigs talking.*
- *I'd like to make a contract about giving myself credit for what I accomplish.*
- *I have a stroke for your difficult work tonight.*
- *No rescues!*
- *I need to discuss a paranoia arising from the last meeting.*

Each of these statements includes a bit of jargon that Group has adopted. The terms *pigs, contracts, strokes, rescue,* and *paranoia* have their origins in Radical Psychiatry, a group therapy movement of the sixties and seventies that focused on the concept of self-empowerment.

The discipline of Radical Psychiatry inspired the formation of

Group. The "very smart and highly gifted therapists" whom Christine credits with helping establish Group were members of the Bay Area Radical Psychiatry Collective, who helped her determine a path to emerge from her depression. It is important to give credit to the influence this movement had on us. This at first seemed an impossible task because I knew of Radical Psychiatry only through Group and as the origin of the terminology we use. But as I reviewed the literature in an effort to describe the core concepts, I came to appreciate the connections between our process and that body of thought.

The simple, powerful message of Radical Psychiatry was that individuals could make changes in their lives by learning to trust and validate their feelings and by making close connections with other individuals. The founders postulated that these connections could be established through problem-solving groups that provide contact (other people one can trust), promote awareness (external input about what is going on), and stimulate action (solutions to individual problems). In such groups, usually led by trained therapists, individuals explored their fears, paranoias, and feelings of alienation. The revolutionary concept was that people could help themselves rather than turning to doctors and therapists for cures. The founders of Group envisioned some key differences, however: it was to be leaderless, it was to consist of people who knew one another and in fact worked together, and it would focus on professional problems. Working in a group to solve individual problems would, in theory, empower the members. Although most of us today don't think of Group as part of Radical Psychiatry, I believe we are a "proof of principle," a success story.

Terminology

Pigs represent "internalized oppression," the judgmental voices of authority that we take on and use against ourselves. I define them as "negative self-perceptions." A pig makes us feel badly about ourselves instead of changing our behavior in a useful way. In Group, we watch for pigs during our own and others' work and try to shake off their influence.

Contracts are concise formulations of objectives. Depending on its context, a contract may state intentions for the next week, the next month, or the foreseeable future. Working contractually is a central principle that allows each of us to refer back to stated objectives and evaluate her progress toward a stated goal. An apparently new problem often relates to a previous contract, allowing us to see connections among various issues. Much of the work in creating a contract is in ensuring that it meets the criteria of being both positive and possible. We avoid "ought" or "should" in contracts, phrasing them as "I will . . ." The contract may be about taking time to think or plan rather than about acting. Contracts may be specific and time limited ("I will draft the discussion section of my paper by the next Group meeting"), specific

but ongoing ("I will remember to ask for what I want"), or cosmic ("I will work toward creating peace in my life"). Some of us
frequently end our work with a specific contract; others use this
format less often. Someone may point out in another's work,
"That sounds like a contract. Do you want to phrase it as one?"
On occasion we record an affirmation, a simple statement to keep
in mind when confronting a problem, instead of or in addition to
making a contract. When someone finds that she hasn't achieved
a contract or isn't making progress with it, the task is to revise it
and learn why it didn't work, not to think that she failed.

A *stroke* is a positive observation about another Group member. Christine called them "little gifts to go home with," because
we share them at the end of the evening. A stroke usually stems
from the meeting—admiration for a subject's work, compli-
menting her attitude, her appearance, or her insights into someone else's work—though sometimes it is generated by something
outside Group. The rules are simple. Strokes should describe attributes of the stroke subject rather than be about the feelings of
the stroker. The person wishing to give a stroke announces her intention and addresses the recipient, who listens and responds with
"thank you," perhaps adding a comment about why the stroke is
important to her. Others sometimes add corollary strokes for the
same person or "a stroke for that stroke" when it seems especially
appropriate or well phrased. A stroke may describe a particular

Representative Contracts and Affirmations

CONTRACTS

- I will select my commitments and then give myself time to do them.
- In the next two weeks I will set up a schedule of regular meetings with each of my postdocs. I will not have more than one meeting in a single day.
- I will exercise and enjoy my authority (the choice of "authority," not "responsibility," is important in this contract).
- I will fully engage in the work I do and will continue to maintain the richness of my life. I will keep to my own center in the light of external input.
- I will give myself credit for incremental steps completed.
- For the enjoyment of my life, I will make my body stronger.
- I will plan my sabbatical to make it a collection of moments instead of a Grand Agenda.
- I will take a one-hour midday break at least two days a week.

AFFIRMATIONS

- I do not have to know this. This is not my field. I can ask advice.
- I am choosing to trust my feelings.
- I have earned the right to set limits to my work, to all parts of my life. I am worthy of my own respect in addition to that of others.

quality—for example, being kind, sensitive, or skillful at putting a thought into words—or a specific action.

It is "bad stroke etiquette" for a recipient to demur with "Oh, but . . ." and explain why she doesn't deserve the stroke. The idea is to listen to what someone else has observed, absorb it, and believe it because of trust and respect for the giver. If the stroke is something that she has a hard time believing about herself, someone may say, "That's hard for me to take in," but the goal is to accept, not protest. This acceptance is surprisingly difficult, especially at first. Helen reported a desire to hide under the table rather than be the focus of all eyes while receiving a compliment. Perseverance pays off, however, because getting a stroke is a delight once one is accustomed to it. I once wished for "stroke sealant" to preserve the good feeling of an especially apt and unexpected stroke.

It is easier to give a stroke than to receive one, and equally delightful. We've all applied the concept of strokes to our external lives too, making an effort to tell our companions, friends, and students positive things we've noticed. We try to recognize when we would like strokes from others, as well as when we know we are unlikely to get them. The best of all possible worlds is being able to stroke yourself, to acknowledge the work you've done.

Avoiding *rescue* is critical to empowering and respecting others. A rescue occurs when we do more than what we want to for someone else or do more than half the work in a relationship.

The rescuer is not being true to his or her desires, and the one who is rescued loses power by being relieved of personal responsibility. In Group meetings, rescue takes two forms. The first is an attempt to solve someone else's problems instead of giving suggestions and comments to help her reach her own solution. This is disempowering and implies that she isn't capable of doing it herself. The second type of rescue is a hesitancy to ask for what we want from Group to protect someone else. Examples of rescue are someone else's deciding not to work on a promotion-related issue because I was sad about not getting tenure, or not asking for assistance because everyone was so busy. We have to ask for what we want and rely on others to take care of themselves. "This is just too hard for me right now" is a valid response to a request.

Identifying the impulse to rescue is often useful in maintaining relationships outside of Group. The problem becomes evident when one of us realizes she has become resentful because she has been neglecting her own needs in favor of the other person's (often a partner or parent). Judith described the result of rescue as "The Fourth Law of Thermodynamics: Rescue is followed by Persecution." Doing what we don't want to *for* someone else inevitably leads to persecuting (e.g., resenting, rejecting, ignoring) that person. Helen phrased it thus: "Put up leads to Fed up."

Paranoias and a companion term, *held feelings*, describe emotions that may interfere with the ability of Group members to work productively together. Such feelings might have arisen from a previous meeting or from conflicts in our relationships with one another outside Group meetings. A paranoia is a fear or concern about how another perceives us. The person who is the object of the paranoia has the responsibility to respond, trying to find the *kernel of truth* in the paranoia. For example, "I felt that you were not interested last session in my work on organizing lab tasks among my students and technicians." The answer might be "I felt really harassed because I need to do that too and I'm so tired and stressed I can't begin to do it now, so I did sort of zone out while you were talking." Finding the kernel of truth is critical, because saying "Oh no, that wasn't so" discounts the other person's perceptions and labels them crazy.

Held feelings are resentments or discomforts with another member of group that we hold but have not admitted to. To deal with held feelings one addresses the person to whom the feeling is directed and gets permission to proceed. (If the recipient feels fragile and unable to hear it, she may say, "No, not now.") The person with the feelings must state clearly what the precipitating event was and how she felt. It is important to make the statement in the proper form: "When you answered the question that Mimi asked me, I felt stupid" rather than "*You made me* feel stupid."

Procedure

Planning and organization are important, both in arranging when to meet and in conducting the meetings. We have been gathering every other Thursday for so long that I was surprised to be reminded in researching old notes that in the early years Group met less frequently, varied the day of the week, and negotiated the next date at the end of each session. With a fixed schedule, people can plan other engagements around meetings. We periodically renegotiate the hour at which meetings start, as we try to balance the demands of working days that run late and the desire to get to bed at a reasonable hour.

The meeting framework is designed to give each member enough time to work and to provide a safe atmosphere. Held feelings or paranoias, if any, are aired at the beginning of a meeting. Next, each member estimates how much time her work will take and signs up for it. This offers practice in asking for what one wants and in making contracts. Agreeing to take a certain amount of time lends discipline as the speaker thinks through what she plans to say. The facilitator keeps track of time, and when it has been used up, asks if the speaker would like more time. If yes, the speaker states how many additional minutes she wants and perhaps says something about what she wants to accomplish. Checking time is the hardest thing about being a facilitator. All of us have trouble interrupting a critical discussion or emotional discovery and are tempted to delay the reminder, hoping for a better time to break in. We try to remain flexible, but discipline about time is critical, because it helps us focus, avoid rambling, and keep our feedback on point, while allowing everyone a turn to work.

It is useful to begin by asking for the kind of response one wants from Group. Our work covers a broad range of subject matter and objectives. We may want, at different times, practical advice, or attentive listening while we share feelings, or perspective on whether we are seeing a situation clearly—is there another way of interpreting events? Someone unsure of where her work is leading may invite interruption: "I'm going to report, but I would be happy to be stopped if it sounds like I'm missing something." If someone finds it hard to ask for what she wants from Group, it may be that she doesn't know what she wants, or because she is by nature reluctant to ask for anything. The process of asking at the start of work is good practice.

Although Group's unique charter is to solve professional problems, our topics have evolved and expanded over time, and there are no rules limiting content. Someone might be trying to identify the source of a diffuse feeling of dissatisfaction. Another might seek help in choosing among possible solutions to a specific problem, for example, the best way to address an employee's poor performance, or the wisdom of protesting an unjust criticism by a boss or grant-review committee. Sometimes we start with specifics and find that our work uncovers a broader theme. I once sought advice on an upcoming professional talk (what topics to include, how to introduce my theme) and ended up in tears about my crowded life, realizing that I was exhausted all the time from moving so fast, even though I thought that only being busy made me feel alive. An individual may also change the focus of her work during the evening, for example, if she finds such strong parallels to her own experience in someone else's work that she decides to address those parallels instead of what she originally planned. Occasionally all the work in a session dovetails, and at the end we all share a single contract for action.

To work on the topics that are most important to us, we each have to trust Group to listen and respect issues. On occasion, one of us has acknowledged that she hesitated to work on a new direction because it wasn't a usual Group topic, only to discover that the rest were very interested. I feared at one time that others would be bored if I worked on being more compassionate toward myself instead of on career goals. Letting other people in on this struggle seemed really hard. Christine encouraged me to take in Group's assurances of interest. "It's like strokes; at first you don't believe them. While you are learning to believe yourself, you have to believe us and if you don't, you have to call us on it. Force us to respond to you in a way that convinces you."

Sometimes a Group member questions her position in Group. In 1995, Helen was concerned that she had lost interest in careers and salary issues since her retirement six years earlier and felt guilty because those issues were still so central to the rest of us. "I don't believe in it any more," she told us. This work developed into a wonderful contract: "I will consider what Group can do for me, ask what is important to me that Group can help with, or can't." Judith noted that each person experiences a natural cycle of greater and lesser involvement in the dominant Group themes and that no one should feel guilt at being in a less involved state. "It is honest. Group is a mirror for intimacy." We built a Group contract: "I have the responsibility to bring to Group the issues that matter to me, and to set aside my judgments of myself."

We have occasionally encountered limitations on our choice of work topics. Because we are friends and colleagues outside of Group, one person's work may cause pain to another. This kind of situation may require a special solution. For example, when one member of Group was responsible for implementing an administrative decision that adversely affected another member, the two were engaged in professional disagreement over some months. Although each could have benefited from using Group as a forum to discuss emotional and practical problems associated with the decision, it was not productive to do so when both were present in Group. The best compromise we could devise was having them agree that the topic was off limits, an imperfect solution because both felt somewhat disenfranchised. Each of them, however, was able to discuss her feelings when the other missed a meeting, or to ask two or three other members to meet as a sub-Group to give advice over a walk or a snack. Meanwhile, the two made special efforts to preserve their friendship, arranging to meet outside of Group to discuss anything *other* than the issue in conflict. In this way they provided each other refuge, a place where life would be free of that one irksome topic.

The Frog Pig and Its Kin

Pigs illustrate how we try to cope with our personal demons and exemplify the Group mix of humor and seriousness that I love so much. We devote considerable energy to identifying, demystifying, and combating the negative perceptions that we call pigs. We identify a pig by certain typical characteristics. One is use of the words *always* or *never* — as in "I always say too much in staff meetings," "I can never write a coherent introduction to a paper," or "I'm incapable of maintaining a long-term relationship." Many pigs are retrospective, using familiar phrases of regret: "I could have . . ."; "I should have . . ." Usually someone who introduces a pig finds that other Group members are familiar with it. We find comfort and wisdom in discerning pigs in common.

Being beset by a pig usually involves losing a sense of kindness toward oneself. The pig says, "I've been incredibly lazy for the last month" instead of the more forgiving "I'm in need of rest and solace because I'm exhausted." A pig says, "I'm clumsy and forgetful" instead of observing with compassion that "I'm so overburdened that I can't function." There is an iterative process we refer to as "pigging myself" for something or "getting all pigged out." One might ask, "Why am I so stressed out? I should be able to handle this. It's good news, I shouldn't be stressed." This statement ignores the fact that even positive change is stressful. Instead of noticing and accepting our feelings, we criticize ourselves for having them. The complaint takes the form "Why am I so messed up that I'm [depressed, frightened, hypersensitive, tired]?" In addition to feeling bad, the victim is criticizing herself for the way she feels.

The frog pig is my favorite. It originated in a discussion of the emotions that come up at a conference when one listens to everyone else promote their research and feels too tired to do likewise. Someone described her fear that everyone would discover her incompetence and called that syndrome the fraud pig. Several of us heard "fraud" as "frog," which became a family joke. Many outside Group recognize the frog pig. Women may be particularly susceptible to this pig, questioning achievements and discounting successes, but friends of both sexes have acknowledged feeling that they have fooled everyone into thinking they are more competent than they really are. One colleague to whom I described this pig was eager to tell her eleven-year-old daughter, "We often talk about those feelings, and it's so convenient to have a name for it."

Being denied tenure was a quintessential setup for my frog pig to attack. I imagined that everyone had suddenly discovered that I was stupid and a bad scientist. My discomfort was so great that I allowed several friendships to atrophy, thereby worsening my professional isolation. After I started my first job outside of academia, and became obsessed with tasks I couldn't finish, Group pointed it out: "You think everyone but you has noticed that you're not doing a good job, that they can see all the things not yet done." And I was amused to find the frog pig lurking again after I quit my corporate job to write, a time at which I pictured myself bravely choosing and following my own path, surely beyond concerns about being exposed as a fraud. Still, I could hear the pig whispering, "You haven't really written anything yet have you? Who are you kidding?"

The frog pig is often accompanied by an urge to tell all. Christine reported on a young student who said admiringly, "You are my role model," on a day when she was feeling down about herself. Her gut instinct was to reply, "Oh yeah? Well, let me tell you . . ." and launch into a tale of her failed relationship, her fantasy of murdering one of her colleagues, and her lack of viable scientific ideas. She had wisely spared the student these reflections, saving them for Group, where they could be appreciated in the right context. If someone says to me, "I think it's wonderful that you are a writer," I want to say, "Well, actually I only worked on the book for five hours last week" or "I'm not sure it will ever be published." I try to take a breath and say instead, "Yes, it *is* wonderful—it's what I've always wanted to do."

Other pigs get names too. Mimi, who does innovative research in diverse areas, said, "There are two pigs for every new scientific endeavor. There is fraud pig at the beginning, when I'm starting out and don't know anything about the field, and has-been pig at the end, when I've done the most exciting things and don't have much more to say." Together, she said, they formed a "moth pig," which characterized her much admired scientific versatility as "flitting from field to field."

The guilt pig finds lots of vulnerable points in our busy lives. When we choose to do one thing, something else always gets neglected, so we feel guilty about the things not done. Someone taking on administrative responsibilities may experience a merciless attack of this pig for neglect of her research. I find the guilt pig most powerful when it is vague and amorphous, so that I don't know where the guilt originates. "If I knew what I felt guilty about I could evaluate it, write it down, stick it on the priority list somewhere—or toss it out. But I let guilt pigs take over without even making them show themselves."

A pig doesn't lose its power just because we have characterized it. Having designated pigs as possibly incapacitating attitudes to watch out for, we then judge ourselves harshly when we fail to identify or combat one. Pigs are tenacious. If we deal with them and find a way to feel better, they will usually counterattack. One member reported that when she introduced more structure into her life because she'd been feeling out of control, the pigs began to accuse her lack of spontaneity. Another made a contract to follow her passion, acting more from the heart than from the head, and the pig voice criticized her for taking an emotional rather than a reasoned approach.

A banished pig may sneakily return, disguised as informed self-criticism. I once wailed, "I examine my behavior, then I use my own insights as further evidence of how badly I'm doing. My pigs grab every insight and I feel worse." I was trying to identify fundamental causes of my disorganization. Instead of saying, "I am tired; I need more rest," the pig phrased it, "This is out of control because I don't get enough rest. My mother always told me that I needed rest. Why haven't I learned yet? If I were stronger, I wouldn't need so much rest. If I didn't waste time, I would have time to rest." And on and on and on . . .

One way to handle a pig is to trap it by writing it down—one of the many things we use lists for. We came up with the idea of a pig notebook, the rule being that "every time you are feeling bad, consider whether there is a pig on hand (there almost always is) and write it down." "Give them numbers!" An extension of this technique is the anti-pig (or auntie pig, a part of the family). After writing down the pig, we try to write an affirmation to counter it; an example of something we can do or have done that contradicts the pig message. For example, "I am a lousy adviser; I never spend enough time with my students" is faced with "Three former students got awards in the last year, and each one mentioned how much I had contributed to his or her development." Pigs *can* be kept at bay, or even permanently banished, but it requires constant vigilance.

When we use visions of pigs to summarize bad feelings, we frequently end up in laughter, which is the best of pig antidotes. I once announced to Group that my pigs were organized. "I think I have unionized pigs." One sunny July we identified the "California fitness pig," which says, "Here I am in California, where the weather is beautiful and everyone knows about the need to exercise, and I'm a complete slug, unable to run 5 miles or do fifty pushups . . . or even sit-ups!" We frequently describe our pigs as "yapping at my heels" or "biting my ankles," absurd embodiments that lighten the load of bad feelings.

Over the years, we have developed a fondness for pig images, which have become a humorous symbol of our struggles. We all have pig collections displayed in our homes—miniature pigs of glass, ceramic, wood, or metal, pig bookends, or plush stuffed pigs. We look for them when we travel and bring them back as presents to other Group members. In contrast to the mental pigs that threaten our well-being, these tangible pigs are a benign species that provide comfort. They remind us to treat ourselves with compassion, the most fundamental of Group precepts.

Some Points to Consider in Establishing a Group

So you are interested in putting a group together. Here are some points to consider. They include objectives and commitments that prospective members should understand, as well as some attributes and interests of likely participants.

- *The central aim is to have members support one another in finding workable solutions to problems.* This is the fundamental defining principle of a problem-solving group as I use the term. Although the statement may seem obvious, it is an important clarification of purpose for people who might otherwise think of "complaining sessions," of "gossip fests," or of an organization committed to broader political goals. *Support, solutions,* and *workable* are all key words.
- *Honesty in presenting one's own issues and in giving feedback is essential.* The heart of this statement is that a group is much less useful if people "make nice" or try to please others, either when presenting problems or in responding to the dilemmas presented by others. Members need to know that honesty is expected of them. Although tension sometimes arises between honesty and caring support, we work to preserve both (see Chapter 17).

  Feedback from a group is especially valuable, because it can be more neutral than input from most other sources. A member can be dedicated to the success of the others, without being personally invested in the specifics of what that success looks like. In this way group participants are more neutral than are family members, and their opinions often more balanced. A partner or spouse will be profoundly affected by the choices one makes and cannot help but be biased as they provide encouragement. A professional colleague may likewise have a stake in an action or decision. The group cares not so much about which solution one chooses as about the rightness of it for the chooser. There is support without dependence on the outcome.

- *It is important to bring in members who are interested in and committed to the process of group problem solving.* This point is best understood by posing the question "Does it sound like a neat idea when you hear about it?" To benefit from a group, a member has to be willing first to acknowledge that a problem exists and second to engage with others in the search for solutions. That may mean hearing unanticipated and divergent points of view about one's own issues. It is necessary to listen, sift, and decide among suggestions. Those who are powerful problem solvers may need practice in accepting a course of action that differs from what she herself has recommended. It's all part of the group process.

- *To create an atmosphere in which people feel safe working on difficult issues, the group should agree upon basic rules of interaction and a framework for running meetings.* Although we have found the rules described in this book very useful, we have also ignored them on occasion. Over time, we have selected the principles that are most appropriate for us, though others might choose a different style. I am convinced that some structure is necessary to guard against power inequities. Each member of a group must have an opportunity to speak and an obligation to listen.

- *The ability to listen is as important as the ability to reason and talk.* Anyone working on a problem should have the chance to express it in her own voice, according to how she sees it. After all, it is her issue. Other group members need to listen and understand that point of view before suggesting alternative approaches. Breaking in with "Of course you feel that way" or "But you should know you are selling yourself short to think that way" is disruptive, though at the proper time such validation or contradiction may be important feedback. Another way of looking at this precept is that some people are very good at persuasive argument and presentation. Some of those who are less skillful speakers may have the gift of really hearing others and catching the implications of their words. Both make important contributions to a group.

• *At the beginning, it is probably important to focus on professional problems. Over time, with experience of one another and the process, group members may or may not choose to include personal issues. The line between personal and professional work, however, is sometimes fuzzy.* This point ties in with the issue of group members developing close friendships. The fuzzy line appears when the demands or concerns of a spouse, child, or companion affect professional struggles, or when there is tension between fulfilling professional obligations and maintaining one's health. Little by little, as trust and respect develop, members may find it easier to talk about nonprofessional issues. As mentioned earlier, some of us were tentative when we wanted to move into more philosophical and emotional work. Listeners need to be honest about their level of comfort as topics become more personal.

• *It is not critical that Group members be close friends.* The women in Group are among my most intimate friends; in many ways they are family. But when I joined Group, I didn't know most of the people in it, and I did not join in the hope of making more friends. I joined to get information, practical help, and emotional support, all of which were forthcoming. Each member of Group has close friends who are not members. Many of these friends and our families validate the work we do in Group through their appreciation of how it helps us. The intimacy that has developed is a tremendous plus but can't be foreseen in starting a group.

- *Setting aside a part of the meeting time for more informal social interaction facilitates the working session.* We coordinate this with our time for strokes and food, which allows us to relax and helps us maintain focus in the work session. Knowing that this period is available, someone who thinks of a troublesome side issue or amusing anecdote during her work may put it aside with "I'll tell you about that during strokes" and continue the central discussion. Likewise, a listener may write herself a note to share an interesting experience later if it isn't directly applicable to the work being done. For a newly formed group, this social time may be an opportunity to test discussion of more intimate issues.
- *The size of a group may vary, but we have found seven or eight to be ideal. Decisions about taking on new members in an existing group must be made by the entire group.* When a group has room for a new member (also a matter for consideration by all), people may seek out and propose likely candidates. It is imperative that all members agree before an invitation is proffered. As we discovered when we were careless about this, someone who encounters a new member without having been consulted feels disenfranchised. A new member can invigorate a group, as everyone reevaluates which

facets of group are most important and considers how best to present the objectives of the group to the newcomer.

- *Members may be from the same department in an academic or a business setting, but they must commit to put competition aside if it exists.* The most relevant help on a specific professional issue usually comes from those who are closely involved with that environment. Our initial group was made up of members from a single institution, and within that, from a small number of closely interacting departments. (It is worth noting, however, that many types of employees were represented—general staff, faculty at various levels, administrators, postdocs. The members felt they were empowered by crucial relationships that were not usually fostered in the bureaucracy.) That worked for us, but a group needs to proceed with caution in the event that there is competition among members, for example, if they are vying for a promotion or job opportunity. In a milieu where weaknesses will be exposed, people clearly must feel confident that their revelations won't be used against them. More subtly, members must guard against feelings of competition, jealousy, or resentment and be willing to expose them and talk them out if they arise.

- *Members need not be in the same discipline. Sharing relevant experiences enhances the function of Group, but people in different walks of life may have issues in common.* This is the flip side of the preceding statement. People with different professional experiences may provide alternative viewpoints. The discovery of like problems in varying environments may combat a sense of isolation even more effectively than learning that everyone else in "your kind of job" has the same problem.

- *The confidentiality of all issues discussed is key, and all members must agree up front to a set of rules governing this.* The underlying goal is that each member must feel free to talk in absolute confidence, knowing that nothing she says will be communicated to anyone outside the group unless by specific discussion and agreement. Members should be able to describe feelings that they find embarrassing, events that are threatening, and proposals for action that are impulsive and not fully considered. The group must be a sounding board, a place to test ideas without constraint; this requires absolute confidentiality.

- *It is important to schedule regular meetings, preferably well in advance so people have a chance to plan ahead to attend.* This rule has a substantial impact on the usefulness of a group. It is more important to have meetings on a regular basis than it is to get a maximum number of people at each meeting. The dynamic that develops when a small sub-group is present can enrich interactions, awaken personal strengths, and assure people that the group's usefulness does not depend on one or two powerful contributors.

These guidelines were formulated from our experiences of the past twenty-five years; some people may prefer a different approach to the group process. For example, a woman in one nascent group told me her fellow members were uncomfortable with signing up formally for individual time at the beginning of each meeting and planned instead to work in a less structured way. Although the makeup and specifics of a group may vary widely and may change over time, an agreement to abide by common rules that mandate support of one another is the foundation of success.

17 *Maintenance and Repair*

WORKING TO KEEP GROUP WORKING

- *Speak the truth but care how you say it. Honesty carries risk of wounding.*
- *Our strength lies in diverse opinions.*

As in any important relationship, it takes work to keep a group going, and serious problems sometimes arise. Despite our commitment to the process and deep concern for one another, we have gone through times when we felt Group was failing or falling apart. I have reacted to these crises with pure terror. A reaction to "fix it at all costs" kicks in, and I just want to make everybody happy and comfortable, even if that means sweeping unpleasantness under the rug. Fortunately, as a Group we've done better than that.

Within any group, conflicts among members are probably inevitable. Each individual strives for completely honest reactions to others' work on fundamental personal issues. The person doing the work, however, may sometimes have difficulty hearing these opinions. There must be mechanisms to allow airing of disagreements and their resolution to guarantee continuation as a productive group. The alternative is a classic rescue, in which people refrain from being honest for fear of hurt feelings.

On occasion, we've held special meetings to discuss how we want Group to function or to fix specific problems. One such session took place at a time when our membership had remained largely unchanged for more than ten years, so that we knew one another well. This incident illustrates the usefulness of Radical Psychiatry procedures and demonstrates that we were so convinced of the value of Group that we were unwilling to let it become something less than it could be.

The issues were fundamental: we wanted to redefine our objectives and develop skills to deal with conflicts among us. We had concluded that we lacked the capability to express disagreement and criticism without hurting others. Several of us found that fear of angering or hurting others was keeping us from working on our own most important issues or from speaking with complete honesty. We wanted to reestablish trust and create some guidelines for preserving it, while being more authentic in our Group meetings, as in our lives. We had attempted to address these problems but had not reached a satisfactory resolution. Feeling that we needed the professional expertise that is lacking in a leaderless group, we invited Beth Roy and Becky Jenkins, the psychologists who helped Group get off the ground initially, to counsel us.

We agreed on a statement of purpose for the meeting: "To learn how to deal with conflict and disagreement constructively and with minimum hurt." Our experts asked us to be prepared to comment on several issues: "What's been positive for me about Group? What's been wrong? List held feelings or paranoias towards others in Group. What do I want Group to be? How can we make it better?" I was strongly in favor of having this meeting, because I was willing to try anything that might help us heal wounds and avoid any further departures. But I was reluctant as well. My notes in answer these questions reveal my conflicted state:

- *I have been resistant to working on this. I don't seem to want to prepare, be there. I can't think of any paranoias or held feelings. I want everything nice, and that doesn't promote honesty. I'd prefer to think of Group against the world, all supportive, celebrating the differences among us. A session dedicated to being better at conflict suggests there will be conflict — how can I feel good about that?*
- *What do I hope to accomplish? I'd like to learn to be more comfortable with conflict within Group, not to feel such intense pain when there is a disagreement.*

I did, in fact, come up with some held feelings and paranoias after all:

- *Held feelings: (1) When I'm giving someone feedback and am interrupted by someone else, I feel resentful. I feel like my timing is off, I'm not heard, and I feel left out. (2) When anyone says something that is clearly hurtful, I feel irritation as well as pain. I'm supportive of people expressing complicated feelings but want them to be gentle. I almost prefer that they not work them out if that will be hurtful to others.*
- *My paranoia: That Group feels my contributions aren't valuable. I fear that I am loved but not respected as a Group member. Maybe I feel inferior because I don't have strong differences with anyone in Group. Thinking about this, I feel cowardly and dishonest.*

As we started the actual meeting, Beth and Becky asked us each to state what we wanted from Group.

- *Group is about problem solving and support. Recently my work has been principally on professional issues. I get lots of support and wisdom. I want to turn to more personal issues, like my father's death. I'm concerned about a lack of honesty here. We need concrete advice on skills to deal with hard, difficult emotions. I want to get off the fix-it role in Group.*

- *Group is a place where each one of us can be authentic, good and bad. Be honest without hurting one another. If we can work through this, we can be more complete people. I see Group as a family.*
- *I want to see a balance between being respectful of others yet honest to them while taking care of ourselves. We need to better understand our differences so we know how to hear each person and consider the consequences of what we say in response. I often feel like an outsider, not heard.*
- *I'm confused about the objectives of Group. I've come to feel an outsider and my alienated feelings make me feel unsafe. I want to find a way to be both real and constructive. I want to establish ground rules and procedures to feel safe . . . and to share things about growing old, and aesthetics.*
- *I was an outsider at the beginning, but now I feel a strong emotional bond to Group. I believe in conflict. I come from a place where it's okay to be angry. Group is a family. I want to not feel bad about raising difficult issues. I want more honesty in the moment, more space for exploring new territory.*
- *I see myself as one of Group's success stories. I have learned to trust women and have developed friendships outside Group that never could have happened without my experience here. My paranoia is that I'm traditional and therefore boring. I wonder if people hesitate to talk to me because I'll judge them. I don't feel responsible for making things better in Group, because I don't know how.*

We told Beth and Becky a bit about the current situation and our concern that we couldn't speak our minds when we disagreed with someone else's actions or analysis. They broke in a few times as we spoke, offering strong opinions about handling conflict. "You need ground rules. It doesn't so much matter what they are, but when shit hits the fan, you go to your stations and follow the rules." They observed that sometimes people feel like outsiders if they don't have any big issues or conflicts with others. They stressed that in a working group the people who are noncontroversial in the sense of loving everyone and having no conflict are extremely important to keeping the group together.

Their guidelines on held feelings and paranoias resonated with me because they showed a consideration for feelings that I had thought lacking in Radical Psychiatry's emphasis on total honesty. A key point was that held feelings should not be stated in the heat of passion. "It's not a venting technique to be used when you just can't keep it in any more. Calm down first. If you are fighting, both of you have to back off." They further clarified this counsel by saying, "It's a mistake to confuse closeness with permission to be cruel. Speaking cruelly to people is *not* a demonstration of love. . . . 'Blowing it out' can be very destructive to a group." There was a caveat: "If the only two choices are blowing it out or silence, the former is preferable if intimacy and honesty are your goals." This advice validated our stated desire to be kind and protective of others and offered guidelines on how to combine honesty with caring.

They reminded us of other guidelines:

- *Don't respond to a held feeling with "I'm sorry." Just say, "I heard it."*
- *Before you respond to a paranoia, make sure you understand it. Other Group members can participate in assuring the understanding. That's the advantage of stating the paranoias in Group rather than in a meeting of two people.*
- *Unvalidated paranoias make people crazy. (I picture this in terms of a loved one slamming doors and talking sharply but, when asked if anything is wrong, saying, "No, I'm fine.")*

They suggested that if a disagreement seemed to be getting out of hand, we should designate a mediator at the moment and agree on a course of action—for example, to use the next meeting to resolve the issue. "Acknowledge the legitimate right of everyone in Group to deal with the argument."

We practiced some techniques. I stated to Christine my paranoia that my contributions to others' work in Group were not seen as valuable. Her validation of the paranoia was that perhaps I sometimes said too much so that the fundamental point or kernel of my feedback got lost. Yet she emphasized that my comments, once clear, were often useful and important. Our counselors pointed out that when paranoia and held resentment are mixed together, neither work. "Anyone is entitled to her resentments if she feels hurt, but stating a paranoia to someone is an *inquiry*, and you have to stick with it as such." One example we worked on that night involved a Group member who felt that another had taken sides against her in a conflict with a mutual friend outside Group. The perceived "taking of sides" generated a lot of anger and resentment, but the *paranoia* was "I fear that you care more about that person than about me," and an appropriately stated response, "I felt that she was more vulnerable than you, so I tried to protect her." Recognizing that we were feeling a little frustrated with how messy these processes are and how complicated it is to get it right, our counselors reminded us, "It's not science. You can leave a conflict not completely fixed. It will come up again." That last statement seemed to me a bit of "good news–bad news," but I appreciated the fundamental message that our relationships could continue in the presence of conflict.

At the end of our session we came up with a Group Stroke for our process. It speaks to our continuing search for honesty and authenticity and acknowledges success: "We've achieved something special by working successfully on difficult life issues while being so close to one another professionally and personally outside of group. Traditionally, a Radical Psychiatry group is made up of people who do not know one another outside of that circle.

Group differed from this model at its inception, having been founded by colleagues from a single institution who wanted to help one another survive in that milieu. Confidentiality and secrecy are much more complex in this type of group."

Group's lessons about friendship are as valuable to me as all the hard, practical work I've done on career choices and professional puzzles. Not only are my relationships with the women of Group among the strongest and most abiding in my life, but the conscious way we tend these relationships affects the way I relate to other friends and family. Group provides a forum for talking out difficulties—and our ground rules demand that we do talk them out.

The demonstration that we can resolve difficulties, even big disagreements, is a powerful lesson for life. In another group, the problems might be different. For example, one or two people might be dominating group discussions, a subset of the members might feel their issues are no longer being addressed, or there could be a real or perceived breach of confidentiality. With or without the help of an outside counselor, a group must define the problem, figure out how to address it, and give each person the opportunity to speak about his or her needs and hopes and disappointments. Since the survival of the group may depend on the resolution, the process is well worth the effort.

Part 4 Epilogue

18 *Another Change of Direction*

LETTING GO AND MOVING ON

The biggest decision I have made with Group's help was to leave my corporate job and dedicate myself to writing. This transition contrasts sharply with my departure from academia for the corporate world fifteen years earlier. Although that change also involved many choices, it was forced on me by a career setback that I did not control. Becoming a writer was my idea. In a way, I was preparing during all the years between the two events, as I learned to honor my dreams and follow my heart. Group was the context in which I could proceed with the right combination of reflection and impulse. This period of my life illustrates many of the conflicts and concerns described in previous chapters and shows how Group helped me deal with them.

A Group meeting at Mimi's house in April 1996 was a critical point. During strokes and goodies we were discussing the idea of a book about us. Our 1994 presentation at the cell biology meetings had generated several inquiries from people interested in writing about Group, but we weren't comfortable with having an outsider tell our story. Christine said, "I was thinking Ellen could write it." And I replied instantly, "You know, I was just thinking that I could . . . if I quit my job."

The idea didn't erupt from nowhere that evening. For some months I had shared with Group fantasies about quitting my job, and my work that very evening had been about retirement, focused on the question "If I were to leave now, would I *care* about [the job of Licensing Director]?" My job was the best I'd had. I admired my boss, she appreciated me, and I had a good staff and terrific colleagues. I was almost satisfied with my compensation. But there was something missing, which I identified as passion for my work. After discussing these feelings in Group, I wrote in my notebook, "I'm not doing the important thing; I'm doing the *other* thing."

I suspected that writing might be my passionate endeavor. I had always written outside the demands of school and jobs, in the form of occasional short stories, poetry, and a journal. In recent months, many of my sporadic notations had been about my desire to write and my inability to make time even for regular journal entries. I felt that if writing had a place in my daily life, "I may one day get to what I want to say." I had decided to keep a journal specifically about retirement plans and fantasies, then lost the notebook I'd chosen, in which I hadn't written a word. Group pointed out that my not writing had deeper roots than misplacing the paper, suggesting the affirmation "It's not about the notebook." Listing reasons why I wanted to find an alternative to my current work, I wrote that I wanted to celebrate life more. "I have a capacity for exultation, but it seems stifled. I want time and space to think about what matters."

My musing went on: "Why is it so terrifying to consider writing? Too many books, too many authors." I had dreamed of writing fiction, but quitting a career to write a novel was too scary, especially as my prose writing had recently been pretty much limited to scientific papers, business memoranda, and letters. Writing a nonfiction piece about Group was a more approachable pursuit. For twenty years I had taken notes not only on my own work but also on hotspots of others' work. The book was a concrete task begging to be accomplished, and I might be particularly suited to it.

I experienced a burst of excited certainty in the weeks immediately following this inspiration. I *knew* I would quit my job to write about Group. This ecstatic confidence dissipated as I began to focus on the complexities and difficulties of completing such a project, pondering how to preserve individuals' privacy while telling our story in a compelling way. I wasn't sure all the Group members were comfortable with my ideas. But by then I had allowed myself to see that I wanted to redirect my life, and it seemed there was no turning back, whether or not I produced "The Book." I described this to Group as getting over the "activation barrier," a phenomenon of physical chemistry that always appealed to me. In many chemical reactions, starting materials are converted to more stable products, but for the reaction to occur, the starting materials need to get extra energy, for example, by being heated. Chemistry texts illustrate this with a graph showing a big hump that represents the energy activation barrier blocking the more stable state. Applied to life, the hump is the kick in the pants that we seem to need to make a change, even when the change is clearly favorable. The idea of writing a book about Group had pushed me to the top of the hump. I now was poised to complete the reaction and change my life, with or without the promise of a successful book.

I didn't quit my job the next day or the next week, although I immediately focused most of my Group work on the details. The project merited lots of thought and discussion. First, there was David. A decision to quit my job would affect his life, for better or worse. His first comment, certainly gratifying, was "But you are so good at it!" He feared that I'd be bored without the stimulation of my job. Then one day that summer, as I was trying to explain why I thought this plan would work for me, he looked at me and said simply, "I think that if you want to retire, you should do it." That was what I needed from him. Later, he became so

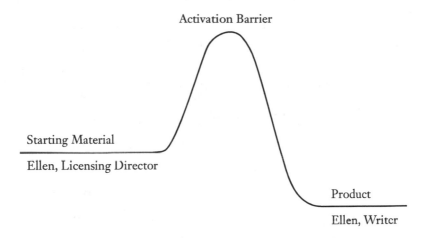

enthusiastic about my writing that I felt a little nervous as he talked of "investing" in my lucrative future writing career.

I was conflicted about finances. It was troubling to think of not making a substantial contribution to our income. I identified things I wanted to be certain I could afford, from travel, theater, and music to financial support of environmental, human rights, and arts organizations. I didn't want my retirement to affect David's ability to retire when he wished. We had "done the numbers," and I was intellectually convinced that we could afford for me to quit, but I was tormented by an impulse to calculate whether I could afford to do it if I had only my personal financial resources. It was as if I couldn't give myself permission to retire if it were possible only because of my supportive husband. I brought this to Group as an attitude I needed to change.

They understood my fixation on finances, recognized that it sprang from long-nourished pride in being independent, and agreed that the calculation would be difficult as well as irrelevant. After working through my confusion, I composed the affirmation "I'm not doing this irrespective of David. My current task is to figure out my career and personal objectives *within* the context of my marriage and my commitment to it."

Group had shared my journey toward deciding to retire and cheered me on without hesitation. But other friends and family heard my idea without having had time to grow into it. A colleague exclaimed, "How can you leave when it's going so well?" It seemed counterintuitive to her, but for me leaving the job on a crest of satisfaction was the best possible time. I was free to quit because I felt successful but had no ambition to advance further. I told one of my brothers what I was considering, perhaps confronting what I imagined would be family disapproval of my giving up well-paid employment at the age of forty-nine. Based on his own experience of retirement, he feared that I would be bored and unfulfilled. On the other hand, we shared memories of the aunt who had retired in her midforties and, as far as we knew, had had no regrets. My brother concluded, "I wouldn't do it, but if it's what you want, go for it!" and my mother commented, "Well, you are lucky to be able to do that . . . but I think you do like your job." I followed up on that lead with Group, going through the familiar exercise of listing the things I liked about my job. It was a long list, but it didn't change my mind. I realized that I could work on ways to make the job less stressful, but the bottom line was that I didn't *want* to make it better; I wanted to do something else.

Who would I be if I weren't working hard, stretching to accommodate workplace demands and outside activities, making a good salary? I had said I didn't want to be defined by my job, but what would be left? On reflection, I decided I was uncomfortable with the lack of action I read into the word *retire* and needed a different word to describe what I was doing. I announced to Group that I'd was going to "re*tread*" instead of "re*tire*."

I explored the feeling that I was not entitled to quit. "Who am I to retire, to do this thing just because I want to?" With Group's help I turned the question around: "But who am I as a human being not to follow my heart when I have the good fortune to be able to do so?" My sense that I needed to earn the right to retire raised a question for all of us. "When does continued success stop meaning additional obligation?" I felt it as a kind of noblesse oblige of the fortunate—I had been so lucky in my life that I owed the world my hard work, and this seemed to have morphed into "hard work at something other than what I want most in life." I wrote, "I want to commit myself to words, to thoughts, to pondering life and love and the world, and I am frightened because it seems self-indulgent and aimless." I wanted reassurance that it was right to take this step, but the degree of rightness had only to do with me, so no one else could reassure me. Organizing my thoughts for Group, I asked myself, practically, "When will I have done enough [to deserve this]?" Writing those words helped me recapture the certainty. "It's not about doing enough; it's about needing to do different things. It's not about having earned the right to change course, to do what I wish (though I believe I have in a sense earned it)—I already have the right!"

Many people reacted to the news that I was retiring by talking about their own thoughts, dreams, and justifications for why and when they might retire (or not). I expected my friend Denise, whom I'd known since we were postdocs together, to disapprove, so telling her was a kind of test of my commitment to the plan. She recognized how her dreams and desires differed from mine. "I wouldn't know what to do with myself if I didn't work in science." As a mother who had balanced work and family, she looked forward to having more time to work as a silver lining to the inevitable sadness of having her daughters leave home. I recognized that Denise, David, and the scientists of Group were alike in a passion for their professions that I had not felt for many years. They all had personal dreams and goals within their careers that competed effectively with the other things they wanted to do when they retired. Recognizing this bolstered my confidence in my choice. It wasn't about my giving up; it was about having different desires. I wrote, "They are figuring out puzzles that matter to them. I need a different puzzle."

I worried that I was deserting people who relied on me. I introduced my staff in the Licensing Department to the concept that I might want to pursue other endeavors without actually telling them I was on a specific trajectory to leave. I obsessed about when and how to communicate with my boss and colleagues. One of my nieces finally said, "Don't overdo this. They won't collapse. People *do* retire. We can't live without you, but they can." This was a big mental exercise for me, who had always worried about taking a single sick day lest it inconvenience someone.

I rehearsed with Group the discussion I would have with my boss. They reminded me not to rescue her by assuming all the responsibility for making my departure smooth and easy. I gave her some options on the details and timing of my leaving, but all the possibilities were comfortable for me. One piece of Group advice was that I should not get into the reasons for leaving unless she asked. "That's not important to her." Always eager to explain myself (and with a tendency to apologize and justify) I had not considered that, and it helped me focus on the purely professional.

In fact, people at work responded with a mixture of interest in me and introspection about their own plans, much like some friends, who were less affected by my decision. Both my boss and a man who had been my most treasured colleague for many years congratulated me, then moved on to discuss the reasons why they were not yet ready to make a similar move. Both had specific things they wanted to accomplish in the technology and business of human health care and understood that I had other priorities. Having prepared with Group not to expect validation from those who would be most affected by my departure, receiving it was a welcome surprise.

As I planned for the future, I talked of many things that I wanted to do in addition to writing. I wanted to spend more time with my mother and do more backpacking, and not always have those activities compete against travel with David. I wanted to enjoy my home, work in the garden, and play the piano, and not be tired all the time. Carol cautioned, "It might not be the job's fault that you are always tired. Don't count on leaving Roche to produce an Ellen who doesn't do too much." Suzanne said, "You have to recognize how much you do for people. Don't leave the job for the wrong reasons." And Christine, "Part of the planning process has to be looking at how busy you are, in all situations." I made a contract to "Take advantage of this special time to see myself fully and appreciate that self."

I wondered whether the things I craved—peace and a slower pace of living, the chance to write, walk, and contemplate— would be less dear if I were not working frantically most of the time. Did I value them so much only because they were restricted, because there was a tension against external obligations? Would I miss contact with people? "Now I anticipate hungrily any amount of alone time—time for the house, the garden, sorting of letters and papers, moments snatched to write in my journal.

But what about the months after the years?" And I was fearful of not having external approval for my achievements. I could rely on Group to applaud my successes, but it would be up to me to tell them what those successes were, to recognize and acknowledge them for myself.

On the other side of the spectrum I worried that maybe I would never accomplish anything: "I'm really frightened of that side that doesn't care whether I have a book to write or not. Seeing how fast the day goes by when I'm not working at my job, I do not believe I could be bored even if I never worked on the book. But maybe when the garden is gorgeous, the photographs are all catalogued, and the storage space is rebuilt, I could wonder what to do."

Right up to the end I had good experiences that made me question my decision, mixed with others that made me delighted to be on the way out. My stepdaughter asked me in June if I felt like a "short-timer" just wanting to leave, wanting to be out of there. I said no, and then found the next day that I absolutely did feel like that, fed up with one person's attitude and irritated by a scolding e-mail from someone else. On the other side, after a productive strategy session with two scientific directors and my favorite attorney, I asked one of them, "How can I do without this?" He said he suspected I would develop the same level of enthusiasm about my new endeavors. When I told my sister-in-law, a scientist, that I loved the job I was about to leave, she said, "But you are a person who would like any job with some intellectual challenge and interest." So I got encouragement. And when I worried that I might not be doing the right thing, I reminded myself, with Group's help, that I was not necessarily making a choice forever. "This is what I choose to do now."

Aftermath

Group threw a wonderful retirement party for me, and there were two more at work. I pushed to get one big project finished before I left, and then suddenly I was no longer a corporate employee. David and I left a few days later on a trip to celebrate my fiftieth birthday. When I returned, I spent three or four days a week at Roche as a consultant to the woman who had taken over my job. This turned out to be a great plan. Knowing I had this grace period, I had been able to depart without leaving everything in perfect shape. With the emotional upheaval of leaving behind me, I could now finish things calmly and with a much clearer eye. I had assured others and myself that I was retiring not because I was burned out but to get on to something more important. After the fact, it became clear to me that I had been burned out but couldn't admit it until I had left the job. I loved watching my successor prosper and was certain that I had done the right thing.

In the first months, I reported "rampant inexplicable joy." I described to Group a morning rearranging the mantelpiece, writing letters, feeling happy to be allowed to clean the garage, hang new pictures. Eventually, I established a rhythm of activity, but at first it was more like floating, being and doing what I wanted, and enjoying it all, relishing the lack of plan. The next phase was overload, a validation of Group's warnings that leaving my job might not resolve my "too much to do" issues. I reported, "I'm passionate about so many things." My hands and arms had begun to hurt when I worked at the computer, and I panicked that something serious was wrong. Group first gave practical suggestions, from keyboard design to B vitamins. Then someone said, "Look at the underlying specter of serious illness—your fear that things won't be okay now that you have made the leap into your dream, that something will interfere." Recalling my past work on living at a pace that allows for contemplation, Group helped me formulate a contract to "Focus on the spaces. Learn to honor the spaces. As they tell you in art books, the spaces are as important as the objects, maybe more so."

I have no qualms about having left my corporate job. Four years later I was delighted to be asked to do a consulting job on university-industry relations. I worked on the project with creative, interesting people, felt that I made a significant contribution, and earned a satisfactory honorarium. But when that project led to the possibility of extended consulting that would have provided more of the same, I declined. I was too jealous of my time, too committed to completing this book, and too compelled by poetry, family, and "other" to want to reintroduce what now felt like my former career into the mix.

Some of the worries I had when I contemplated retirement have materialized. I do struggle to find the time to write enough and trash myself when I don't. I developed the contract "I choose to introduce discipline into my life in a way that fosters creativity" to help me focus on what my writing projects. I also sometimes miss the clear milestones and the external approval that have been replaced by frequent "How's the book coming?" from friends and family. But I have not found, as I once feared, that the things I want to do are less satisfying now that I have to balance only among them, whereas before I was balancing all of them against the demands of a job. Never once have I been bored or regretted my move. And there have been surprises. Even as I dedicated myself to finishing this book, I found that my retirement had made room for time with my granddaughter, whose parents moved to California shortly after she was born. She is a worthy competitor (as is her sister, who arrived two years later) for my writing hours.

In all these areas, I have work to do with Group, and I continue to get valuable insights, challenges to my viewpoint, and great pleasure from the meetings that take place every other Thursday.

Biographies of Group Members

The members of Group are listed in the order in which they joined
Group.

Christine Guthrie (1977) is professor of biochemistry at the University
of California, San Francisco (UCSF). Born in Brooklyn, New York, and
educated at the University of Michigan (BS in zoology) and the University of Wisconsin (PhD in genetics), she joined the faculty at UCSF in
1973. The hallmark of her research is the use of genetics to understand
molecular mechanisms regulating gene expression. Her current work
focuses on the splicing and export of the messenger RNA (mRNA) of
eukaryotic (animal) cells. The mRNA copy of a DNA sequence (gene)
contains meaningless "junk" information that must be precisely removed
("spliced") to make a functional mRNA that provides the cell directions
for making the protein that the gene encodes. Christine is a member of
the National Academy of Sciences and the American Academy of Arts and
Sciences and is a lifetime professor of the American Cancer Society. She
has long believed that the best science happens in a nurturing environ-

ment and in 1998 received the Women in Cell Biology Senior Career
Recognition Award of the American Society for Cell Biology in honor
of this practice. She has trained forty-seven graduate students and postdoctoral fellows, who have gone on to many institutions, including faculty
positions at UCSF, Harvard University, and the University of Arizona.
Christine is married to John Abelson, an emeritus professor of biology at
the California Institute of Technology. They (finally) live together in a
home they recently built in San Francisco.

Ellen Daniell (1977) is a writer living in Oakland, California. Born
in New Haven, Connecticut, she received her BA from Swarthmore
College and her PhD from the University of California, San Diego, both
in chemistry. She spent eight years as assistant professor of molecular
biology at the University of California, Berkeley. Moving to the biotechnology industry, she held management positions in human resources and
patent licensing at Cetus Corporation and was Director of Licensing at
Roche Molecular Systems. She always retained a great personal interest
in science and science education and retired from the corporate world
to pursue a writing project that became this book. Ellen is married to
David Gelfand, a scientist in biochemical genetics, and is a proud step-
mother of one. With the birth of their two granddaughters, she has been
astonished to learn she has a talent for, and takes great delight in, being
a grandmother.

Helen Wittmer (1980), now semiretired, is an active grandmother
in Berkeley, California. She was born Helen Marie Felipe in Oroville,
California, her parents having come to the United States from northern
Spain. She moved to the Bay Area to attend business college and after
graduation worked at an insurance company, at the public radio station
KPFA, and for the Berkeley School district. A resident of Berkeley for
fifty-seven years, Helen loves the mix of people, ideas, culture, and
politics that thrive there. In 1971, she took a job in the Department of

Molecular Biology at the University of California, Berkeley (UCB),
where she worked for nineteen years, first as a receptionist, then as a
graduate secretary, and finally as an administrative assistant to the department chair. These jobs began her association with scientists, academic
science and, ultimately, Group. Since her retirement from UCB, she has
transcribed oral historics at California State University, Sacramento, and
has worked part-time as an office assistant to two Berkeley psychiatrists.
Helen/Wittmer has two children and two grandchildren who live nearby.
Helen's thought for readers of this book: "I believe it would be a far more
humane world with women in charge."

Suzanne P. McKee (1980) is a senior scientist at Smith-Kettlewell Eye
Research Institute in San Francisco. Smith-Kettlewell is a private, nonprofit organization dedicated to the study of all aspects of human visual
processing. A visual neuroscientist, Suzanne has researched the properties
of human vision for more than thirty years; her special interests are
human binocularity and stereopsis (three-dimensional vision.) Born in
Vallejo, California, Suzanne is a third-generation Californian; she left
the state for the first time to attend Vassar College. After completing her
undergraduate degree, she returned to the University of California,
Berkeley, where she earned a PhD in experimental psychology. Her
mentor at Berkeley was Professor Gerald Westheimer. Suzanne is
a former vice-president and trustee of the Association for Research in
Vision and Ophthalmology and a fellow of the Optical Society of America.
She is a member of the editorial boards of *Vision Research, Journal of Vision,*
and *Ophthalmic and Physiological Optics,* in addition to being an executive
editor of the journal *Perception.* Suzanne has trained twelve students and
postdoctoral fellows who are now in research and faculty positions at
institutions including Brown, Rutgers, and Oxford universities. Currently,
she serves on the study section of the National Eye Institute, covering
topics in central visual processing. She lives in Berkeley with a charming,

somewhat befuddled cat and her husband of forty years, Christopher F. McKee, a professor of physics at the University of California. She has three children and two grandchildren.

Mimi A. R. Koehl (1981) is professor of integrative biology at the University of California, Berkeley. She studies the physics of the interaction of organisms with their environments, focusing on such issues as how microscopic creatures swim and capture their food, how wave-battered marine plants and animals avoid being washed away, how olfactory antennae catch odors from the air or water around them, how flight evolved, and how the "skeletal" design of squishy animals like worms and sea anemones came to be. She loves the out-of-doors and does some of her research in the field, on coral reefs and wave-swept coasts around the world. A native of Washington, D.C., Mimi earned her BA in biology from Gettysburg College (having started out as an art major) and has an abiding interest in the intersection of art and science. She earned her PhD in zoology from Duke University, and after postdoctoral work at Friday Harbor Laboratories (University of Washington) and in England (University of York), she served on the faculty at Brown University for a year before moving to Berkeley. Mimi was awarded a Guggenheim Fellowship in 1988, a MacArthur Grant in 1990, and the Borelli Award in 2002 for "outstanding career accomplishment" in biomechanics. She is a member of the National Academy of Science and the American Academy of Arts and Sciences. She has trained thirty-two graduate students and postdoctoral fellows, who have gone on to faculty positions at University of California, University of Oregon, Oregon State, the University of North Carolina, and Bowdoin College. Mimi lives in Berkeley with her husband, Zack Powell, an oceanographer and professor at Berkeley. She has two stepdaughters and four grandchildren.

Judith P. Klinman (1981) is professor of chemistry and of molecular and cell biology at the University of California, Berkeley, and a former chair of the Department of Chemistry. She studies how enzymes, the body's catalysts, speed up reactions by as much as ten to the twentieth power. Along the way, she has discovered a new class of enzyme-bound cofactors and demonstrated the importance of quantum chemistry when enzymes break carbon-hydrogen bonds. Her work has also shown how enzymes have evolved to make use of oxygen while avoiding its toxic by-products. Judith was born in Philadelphia, educated at the University of Pennsylvania (AB and PhD) and was a research scientist at Fox Chase Cancer Center for ten years before moving to Berkeley. She has trained seventy-two graduate and postgraduate students who have moved on to a range of academic and nonacademic endeavors, among them positions at Johns Hopkins University, the University of Massachusetts, Glaxo Smith Kline, Cold Spring Harbor Laboratories, Merck and Company, and *Nature* magazine. Judith has been elected to the National Academy of Sciences, the American Academy of Arts and Sciences, and the American Philosophical Society. She is greatly appreciative of her four children (two sons, a stepson, and a stepdaughter), who tolerated her obsession with science and sometimes absent-minded presence, and her three beautiful grandsons. Judith divides her time between Berkeley and Sonoma County, where she and her longtime partner, Mordecai Mitnick, share the pleasures of good friends, food, wine, and the beauty of the countryside.

   Carol Gross (1993) is a professor in the Departments of Microbiology and Immunology and Cell and Tissue Biology at the University of California, San Francisco. Born in Brooklyn, New York, she received her BS from Cornell, her MS from Brooklyn College, and her PhD from the University of Oregon. Before moving to San Francisco, Carol was first a research professor and then a faculty member at the University of Wisconsin (Madison). She studies regulatory networks that govern what proteins cells make in response to environmental change. Her experimental system is the bacterium *Escherichia coli*. Carol is a member of the National

Academy of Sciences, the American Academy of Arts and Sciences, and a fellow of the American Association for the Advancement of Science. She has trained forty-three students and postdocs, whose subsequent appointments include the National Institutes of Health, Tufts University, the Massachusetts Institute of Technology, Texas A & M, the University of California, Davis, and, in industry, Smith-Kline, Avida, and Dupont. Carol is dedicated to recruiting underrepresented groups to science and to helping all students succeed by improving mentoring. In honor of these efforts, the American Association for the Advancement of Science chose her to receive the Mentor Award in 1993. Carol has two children, four stepchildren, and six grandchildren.

Notes

Introduction

1. D. J. Nelson and J. C. Rice, *A National Analysis of Diversity in Science and Engineering Faculties at Research Universities* (Norman, Okla., 2004), http://cheminfo.chem.ou.edu/faculty/djn/diversity/top50.html.

2. G. Evans, *Play like a Man, Win like a Woman: What Men Know about Success That Women Need to Learn* (New York: Broadway Books, 2001).

3. *A Study on the Status of Women Faculty in Science at MIT* (Cambridge: Massachusetts Institute of Technology, 1999), http://web.mit.edu/fnl/women/women.html.

4. National Research Council, *From Scarcity to Visibility: Gender Differences in the Careers of Doctoral Scientists and Engineers* (Washington, D.C.: National Academy of Sciences Press, 2001), http://www.nap.edu/books/NI000366/html.

5. Elizabeth Zubritzky, "Women in Analytical Chemistry Speak Out," *Analytical Chemistry* (1 April 2000): 10.

6. Lawrence H. Summers, "Remarks at NBER Conference on Diversifying the Science and Engineering Workforce," http://president .harvard.edu/speeches/2005/nber.html.
7. Andrew Lawler, "Summers's Comments Draw Attention to Gender, Racial Gaps," *Science* 307 (2005): 492; Yu Xie and Kimberlee A. Shauman, *Women in Science* (Cambridge: Harvard University Press, 2003).
8. Summers, "NBER Conference"; Lawrence H. Summers, "Letter to the Faculty Regarding NBER Remarks," http://president.harvard .edu/speeches/2005/facletter.html.
9. Marcella Bombardieri, "Summers' Remarks on Women Draw Fire," *Boston Globe*, 17 January 2005.
10. Piper Fogg, "Harvard's President Wonders Aloud about Women in Science and Math," *Chronicle of Higher Education* 51, no. 21 (28 January 2005): A12; Carol Muller et al., "Gender Differences and Performance in Science," letter to the editor, *Science* 307 (18 February 2005): 1043.

2 Evolution

1. "Women in Cell Biology Committee. A Women's Professional Problem-Solving Group"; an audio presentation of this talk is available on the Web site of the American Society of Cell Biology, http://www .ascb.org/audio/94/wicb94summary.html.
2. Radical Psychiatry and its relation to Group are explained in Chapter 16.

4 Accepting . . . Liking . . . Celebrating

1. J. K. Conway, *When Memory Speaks* (New York: Alfred A. Knopf, 1998).

5 A Serious Mind and a Light Heart

1. D. G. Myers, *Intuition: Its Powers and Its Perils* (New Haven: Yale University Press, 2002).

Notes

14 *Anticipating Changes*

1. Pema Chödrön, *When Things Fall Apart: Heart Advice for Difficult Times* (Berkeley: Shambhala Library, 1997).

Further Reading

Groups and Radical Psychiatry

Roy, B., and C. Steiner, eds. "Radical Psychiatry: The Second Decade." Manuscript, 1994. Available from Beth Roy, 270 Prospect Street, San Francisco, Calif. 94110. Excerpts on-line at http://www.emotional -literacy.com/rpo.htm.

Steiner, C. *Scripts People Live: Transactional Analysis of Life Scripts.* 2d ed. New York: Grove Weidenfield, 1974, rpt. 1990.

Mentoring

Fedele, Nicolina M. , and Elizabeth A. Harrington. *Women's Groups: How Connections Heal.* Wellesley Centers for Women, No. 47 (Wellesley, Mass.: Wellesley College, 1990), http://www.wcwonline.org.

Further Reading

MentorNet. The E-Mentoring Network for Women in Engineering and Science, http://www.mentornet.net.

Wellesley Centers for Women Web site., http://www.wcwonline.org.

Status of Women in Science and Engineering

Committee on Women in Science and Engineering (CWSE), http://www7 .nationalacademies.org/cwse. A standing committee of the National Research Council whose mandate is to coordinate, monitor, and advocate action to increase the participation of women in science and engineering.

Fogg, Piper. "Harvard's President Wonders Aloud about Women in Science and Math." *Chronicle of Higher Education* 51, no. 21 (28 January 2005): A12, http://chronicle.com.

Gender Differences in Careers of Science, Engineering, and Mathematics Faculty, http://www7.nationalacademies.org/cwse/Gender_differences.html. A project of the National Research Council, begun in 2004 and sponsored by the National Science Foundation to examine issues such as faculty hiring, promotion, tenure, and allocation of institutional resources, including laboratory space. The expert committee that will carry out this project is drawn from academia and industry to represent a broad spectrum of experience and viewpoints. Reports of gender and diversity studies carried out at many universities are also available through this link. The National Academies track ethnic and racial diversity as well.

Gender Matters: Women and Yale in Its Third Century, http://www.yale.edu/ wff/gendermatters/. Proceedings of a forum held in conjunction with Yale's tercentennial celebrations.

Günther, Donna K. "Why Women Earn Less: Economic Explanations for the Gender Salary Gap in Science." *AWIS* 33, no. 1 (2004): 6–10.

Lawler, Andrew. "Summers's Comments Draw Attention to Gender, Racial Gaps." *Science* 307 (2005): 492.

Rapoport, Alan L., dir. *Gender Differences in the Careers of Academic Scientists and Engineers: A Literature Review.* Special Report, NSF 03-322. Arlington, Va.: National Science Foundation, 2003.

Women in Biology Internet Launch Pages, http://www-rcf.usc.edu/~forsburg/bio.html. This site provides a starting point for finding information for biologists who happen to be women. It is a list of bookmarks to the ample original content already available on the Web. Many of the links are aimed toward women who are graduate students, postdocs, or more senior scientists, but there are also sites relevant to undergraduates or even high school students who may be contemplating a career in biology. These are intended to help women biologists with the practical aspects of busy professional lives and to provide food for thought in those quieter moments. You will find the history of women in science, aspects of science education, an extensive list of career resources for PhDs, postdocs, and beyond, and, of course, information about the specific challenges women face in a sometimes chilly climate.

Yu Xie and Kimberlee A. Shauman. *Women in Science: Career Processes and Outcomes.* Cambridge: Harvard University Press, 2003.

Other

Miner, Valerie, and Helen E. Longino. *Competition: A Feminist Taboo?* New York: Feminist Press of the City University of New York, 1987.

Acknowledgments

Numerous people read the manuscript and gave me suggestions and advice at various stages of its preparation. Early on, Ruth McNeill provided a "nonscientist" view, and my sister-in-law Laurie Daniell, who, sadly, never saw the final product, gave me her unique "scientist-in-the-family" view. They liked my work for different reasons, offered opposing suggestions, and encouraged me to keep writing. I am indebted to Jean McMann and the participants of her Craft and Inspiration writing workshop, and to Annie Barrows, my first professional editor, who showed me where my writing was strongest and why. Annie's prodding to make the book accessible to a wider audience forced me to find a balance between my unique experience and the broad applicability of group work.

The members of Group inspired me to write, commented generously and honestly on content and style, and reminded me again and again to write from my heart. Ben Roberts guided me through the maze of the publishing world and kept assuring me that I would find a publisher, as he helped me do so. This leads to Jean Thomson Black, my editor at Yale University Press. I thank Jean not only for her labors in the publication process but also for her belief that the subject was important and should be published. I am grateful to Beth Roy, Joan Steitz, Liz Blackburn, Nancy Hopkins, Donna Shalala, Linda Wallace, Sheila Patek, Harold Varmus, Tom Cech, Shirley Tilghman, Bruce Alberts, Joan Bennett, Rita Colwell, Keith Yamamoto, and other reviewers for thoughtful comments and critical suggestions. The bursts of enthusiasm from many of you kept me on track prior to acceptance of the manuscript for publication. Laura Davulis, Karen Gangel, and other staff at Yale University Press made the process smoother than this novice author could have hoped. Last but by no means least, I thank my husband, David, who, *comme d'habitude*, encouraged me to do whatever I thought best, had confidence that I'd succeed, and was patient with the process.

KIN KONG

Ellen Daniell is an author and consultant. She was assistant professor of molecular biology at the University of California, Berkeley, and has held management positions in human resources and patent licensing in the biotechnology industry. She has spoken to various organizations on the issues addressed in *Every Other Thursday*, including the following groups:

UC Davis School of Medicine
Memorial Sloan-Kettering Cancer Center
Business and Professional Women's Foundation
    (keynote speaker, fiftieth anniversary celebration)
Women's Global Health Scholars Program,
    University of California, San Francisco
University of Zurich
Swarthmore College
University of Washington
Yale Women Faculty Forum